Cometary Encounters

Cometary Encounters

Flash-Frozen Mammoths, Mars–Earth Discharge, Comet Venus and the 3,600-Year Cometary Cycle

Pierre Lescaudron

Red Pill Press
2021

Copyright © 2021 Quantum Future Group Inc.

ISBN 978-1-7349074-2-1

Published by Red Pill Press (www.redpillpress.com)

First edition.

No part of this publication may be reproduced, stored in a retrieval system, or transmitted in any form or by any means, electronic, mechanical, or otherwise, other than for "fair use," without the written consent of the author.

Contents

Acknowledgements	i
Introduction	iii
Part I: Of Flash-Frozen Mammoths and Cosmic Catastrophes	1
The Woolly Mammoths	3
The Younger Dryas	7
The Crime Scene	11
The 'Event'	17
What Is Flash-Freezing?	21
Recorded Cases of Sudden Cooling on Our Planet	25
Atmospheric Ablation Induced by a Cometary Impact	29
Wandering Geographic Poles	37
Location of Geographic North Pole before Impact	41
Crustal Slippage	45
The Coroner's Verdict	49
The Tragic Fate of the Woolly Mammoths	53
Mega Tsunamis	59
No Human Remains But Archaeological Evidence	61
Conclusion	65

Part II: Did Earth 'Steal' Martian Water?   67

Sea Levels on Earth   69

Water on Mars?   71

How Could Mars Lose Its Water?   73

Interplanetary Electric Discharge   77

Relative Polarity of Mars and Earth   83

Signs of Electric Discharge on Mars   85

Evidence of Material Transfer from Mars to Earth   89

How Could Mars Get So Close to Earth?   97

Was Venus a Comet?   99

When Did the Water Transfer Occur?   105

Conclusion   113

Part III: Volcanoes, Earthquakes and the 3,600-Year Comet Cycle   117

The 14,400 BP (12,400 BC) Event   119

Cometary Cycle?   127

The 3,600 BP (1,600 BC) Event   131

The 7,200 BP (5,200 BC) Event   139

The 10,800 BP (8,800 BC) Event   147

Mystery Eruptions   155

Correlation between Cometary Activity and Volcanic Activity   165

Conclusion   173

Part IV: The Seven Destructive Earth Passes of Comet Venus   177

Venus Markers   179

Looking for a Date	183
Methane Spike and Temperature Drop	185
Deuterium	191
Sulfur Dioxide	193
Carbon Dioxide	195
Increased Atmospheric Dust	197
Increased Wetness	199
Impact on Human Populations	205
Cometary Venus in Myth	211
Zooming in 5,200–4,600 BP	219
Conclusion	227
Epilogue	229
Bibliography	237
Index	253

Acknowledgements

Even if there is just one name on the cover, a book is rarely an individual endeavor. While writing this book I saw how important the help from others was and I would like very much for you to know who helped transforming what were vague ideas a few years ago into the tangible reality you're holding in your hands right now.

My thanks go to all the great people who closely or remotely provided their help along the way. I'd like especially to thank the members of the FOTCM[1] community in France with whom I live and entertain daily conversations, which contributed immensely to the ripening of the ideas exposed below.

My gratitude also goes to the editors and contributor of the *Sott.net*[2] website and the *Cassiopaea* online forum for gathering and sharing tons of invaluable information which provided great inspiration for the present book.

Last but not least, I want to say a warm thank you to Laura Knight-Jadczyk for her numerous insights and support, to Joe Quinn and Harrison Koehli for painstakingly correcting my broken English, to Damian Assels for designing the atmospheric ablation diagram, to Chad Seabrook for providing a number of enlightening articles and to Myriam Kieffer for her skillful cover design.

---

[1] Fellowship of the Cosmic Mind.
[2] Signs of the Times.

Introduction

I finalized writing my first book, *Earth Changes and the Human-Cosmic Connection*, in 2014.[3] I believed it would be my last. However, during the research pertaining to that book, in particular the chapters related to cometary impacts and global cooling, a question kept nagging me: How could massive and robust creatures like woolly mammoths become frozen in a matter of minutes?

I began researching this topic in 2015. Initially it was supposed to be just an article to provide some tentative explanations for the peculiar demise of the woolly mammoths. Nothing more.

But, as is often the case, the writing of this first article titled "Of Flash Frozen Mammoths and Cosmic Catastrophes,"[4] which constitutes the first part of this book, brought more questions than answers. In particular it revealed an oddity: the sudden death of the mammoths happened during a severe cooling episode known as the Younger Dryas. Such a cooling should increase the volume of polar ice and, as a result, reduce sea levels. However, during the Younger Dryas, sea levels rose 17 meters over more than a millennium.

What could explain this apparent paradox? That is this central question that led to a second article titled "Did Earth 'Steal' Martian Water?"[5] which constitutes the second part of this book.

The rabbit hole kept getting deeper. The demise of the mammoth (ca. 12,900 BP) and the Earth-Mars interaction (ca. 12,600 BP) were only two of the three major cooling events that led to the Younger Dryas. There was a third sudden cooling ca. 14,400 BP. What could be its cause?

Interestingly, events similar to the 14,400 BP event seem to have also occurred in 10,800 BP, 7,200 BP and 3,600 BP, leading to an hypothesized 3,600-year cometary cycle, hence the title of the article "Volcanoes, Earthquakes And The 3,600 Year Comet Cycle,"[6] which is the third part of this book.

While writing the second article about Mars' close encounter with Earth, I suggested – in line with some of Velikovsky's theories – that Mars was pushed close to Earth by Venus, which, at the time, was a cometary body. While I focused on an Earth-Mars interaction, a central question was left unanswered: "What happened to Venus after it pushed Mars off its orbit?" The answers to the question were given in an article titled "The Seven Destructive Earth Passes of Comet Venus,"[7] which constitutes the fourth part of the present book.

---

[3]Written in English and published in May 2014 by Red Pill Press. Translated in French, German and Russian.
   Available on Amazon: https://www.amazon.com/Earth-Changes-Human-Cosmic-Connection/dp/1897244975
[4]Published in 2017 on sott.net.
   https://www.sott.net/article/357709-Of-Flash-Frozen-Mammoths-and-Cosmic-Catastrophes
[5]Published in 2019 on sott.net.
   https://www.sott.net/article/420386-Did-Earth-Steal-Martian-Water
[6]Published in 2019 on sott.net.
   https://www.sott.net/article/424968-Volcanoes-Earthquakes-And-The-3600-Year-Comet-Cycle
[7]Published in 2020 on sott.net.
   https://www.sott.net/article/432498-The-Seven-Destructive-Earth-Passes-of-Comet-Venus

So, here we are, five years after first questioning the nature of the woolly mammoths' demise, addressing topics that have little or nothing to do with mammoths, but which show the frequency and severity of cometary events and raise a question that in a roundabout way brings us back to the initial topic: "Will living creatures in general and humans in particular have the same tragic fate as the woolly mammoths?"

This book is not the mere compilation of four existing articles. It offers enriched footnotes, more than 750 in total, improved illustrations, almost 200, some featuring extra details, updated and augmented text including extra chapters – and an extensive bibliography listing more than 300 papers and books. All in all, the content has been increased by more than 50%, but the above only deal with quantitative factors. What really matter are the qualitative ones and they are for you to judge. So, let's get started!

Part I:
Of Flash-Frozen Mammoths and Cosmic Catastrophes

Figure 1: © Comet Research Group
Artistic depiction of the cometary-induced demise of the woolly mammoths

For years I've been fascinated by one of the greatest mysteries of our planet: the demise of the woolly mammoths. Try to imagine the barely imaginable: millions of giant creatures inexplicably flash-frozen overnight.[8]

This is a fascinating event for several reasons. First, flash-freezing is a very peculiar process that does not really occur on our planet. Also, given the circumstances, the magnitude and power necessary to virtually wipe out the whole mammoth genus is truly astounding. But maybe the most fascinating aspect of this event is that it occurred just 13,000 years ago when the human race was already widely established on planet Earth. For comparison, the Magdalenian[9] cave paintings found in Southern France[10] were made 17,000 to 12,000 years ago.

This event challenges our uniformitarian vision of history where the progress of life on our planet is a linear process, increasing day after day, undisturbed by any major external setback. Therefore, such an event casts a different light on our human condition and the pervasive delusion that human beings are somehow above natural laws, including those that govern major catastrophes.

It is a fascinating and puzzling topic because the numerous theories that have been proposed over the last two centuries to explain the demise of the woolly mammoths - such as them being caught in frozen rivers, victims of over-hunting, covered by hail storms, buried in mudslides, fallen into ice crevasses, caught by the ice age – are not sufficient to **explain this mass extinction and all its features**.

Figure 2: © Michelin
Mammoth painting in Rouffignac cave

So, in the following we will try to find explanations about how and why millions of woolly mammoths ended up flash-frozen overnight.

---

[8] Along a 600-mile stretch of the Arctic coast were found more than half a million tons of mammoth tusks. Because the typical tusk weighs 100 pounds, this implies that about 5 million mammoths lived in this small region. See: Mark Krzos (2006), "Frozen Mammoths," p. 12.

[9] The Magdalenian cultures span from around 17,000 to 12,000 BP (Before Present). It is the latest development of the Upper Paleolithic, which approximately covers the 50,000–10,000 BP time period.

[10] Those painting are found in dozens of caves like Lascaux and Rouffignac in Dordogne, Niaux in Ariège, Pech Merle in Lot, Chauvet in Ardèche.

The Woolly Mammoths

The woolly mammoth is a close cousin of the modern elephant. Its size was similar to the African elephant, males reaching a shoulder height of about 3 meters[11] and weighing up to 6 tons.[12] The mammoths had a plant-based diet and a full-grown male would need to **eat between 60 and 300 kg[13] of plants daily.**[14]

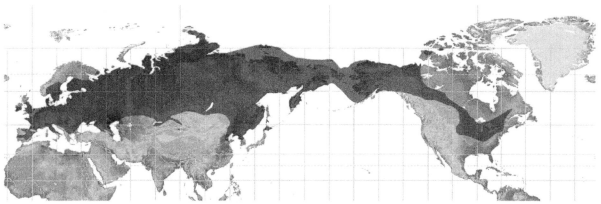

Figure 3: © Kahlke
Maximum extension of the woolly mammoth during the Late Pleistocene.

At the time, woolly mammoths were very abundant on our planet. To illustrate this point, between 1750 and 1917, trade in mammoth ivory prospered over a wide geographical region, yielding an estimated 96,000 mammoth tusks.[15] It is estimated that about 5 million mammoths lived in just a small portion of northern Siberia.

Before their extinction, woolly mammoths inhabited wide portions of our planet, as indicated by the dark grey area in the map above. Mammoth remains have been found all over Northern Europe, Northern Asia and Northern America.

Woolly mammoths were not the new kid on the block either; they had been roaming the planet for six million years[16] before modern elephants and woolly mammoths split into separate species. A prejudiced interpretation of the hairy and fatty nature of the creature, and a belief in unchanging climatic conditions, led scientists to deem the woolly mammoth a creature of cold areas of our planet. But furry animals don't necessarily live in a cold climate – see, for example, desert animals like camels, kangaroos and fennecs. They are furry and they live in hot or temperate climates. In fact, most furry animals could not survive arctic weather.

---

[11] 10 feet.
[12] Alina Bradford (2016), "Facts About Woolly Mammoths," *Live Science*.
[13] Between 130 and 660 pounds.
[14] Berkeley Editors (2020), "About Mammoths," *University of California in Berkeley*.
[15] Llan Starkweather (2007), "Earth Without Polarity," *Lulu.com*, p. 24.
[16] Brian Handwerk (2005), "Woolly Mammoth DNA Reveals Elephant Family Tree", *National Geographic*.

What makes for successful cold adaptation is not fur *per se* but its erectile nature,[17] which traps a layer of air for thermal insulation against the cold. Unlike the forest hog[18] for example, mammoths were devoid of erectile fur.[19]

Another factor in the protection of animals against damp and cold is the presence of sebaceous glands, which secrete oil on the skin and fur. Being water repellent, oil offers protection against dampness. The woolly mammoth had no sebaceous glands[20] and its dry hair would have allowed much snow to touch the skin, melt,[21] and dramatically increase heat loss.

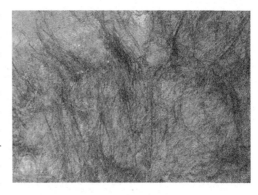

Figure 4: © Naturhistorisches Museum Wien
Woolly mammoth "wool"

As suggested by the picture above, mammoth hair was not very dense. For comparison, the fur of the yak (a cold-adapted Himalayan mammal) is about 10 times thicker.[22] In addition, the mammoths had hair on their toes,[23] yet every arctic animal has *fur*, not hair, on their toes. Hair would have caused snow to cake on its ankles and hinder walking.

The above clearly shows that hair is not proof of cold-adaptation, and neither is fat. Fat only proves that food is plentiful. A fat, overfed dog could not withstand an arctic blizzard and its -60°C[24] temperatures. On the contrary, creatures like arctic hares[25] or caribous can, despite their relative low fat-to-body-mass ratio.

Figure 5: © Cool Antarctica
The lean and cold-adapted arctic hare

Also, mammoth remains are usually found piled up with other animals,[26] like tiger, antelope, camel, horse, reindeer, giant beaver, giant ox, musk sheep, musk ox, donkey, badger, ibex, woolly rhinoceros, fox, giant bison, lynx, leopard, wolverine, hare, lion, elk, giant wolf, ground squirrel, cave hyena, bear, and many types of birds. Most of those animals could not survive the arctic climate. This is an extra indication that woolly

---

[17] American Physical Society's Division of Fluid Dynamics (2015), "How does fur keep animals warm in cold water?" *ScienceDaily*.
[18] Jean D'Huart *et al.* (2013), "Hylochoerus meinertzhageni – Forest Hog," *Mammals of Africa: Volume VI*, pp. 42–49.
[19] David Hopkins, *et al.* (1982), "Paleoecology of Beringia," *New York Academic Press*, p. 285.
[20] Ibid.
[21] As illustrated by igloos, snow exhibits low thermal conductivity, but becomes highly conductive once it melts. The thermal conductivity of liquid water is about twelve times higher than the conductivity of snow.
[22] Kneyda Spins Editor (2013), "Woolly Mammoth Wool?" *Kneyda Spins*.
[23] Avery, Appiah, (2008), "Mammuthus primigenius – The Woolly Mammoth," *McMaster University*.
[24] -80°F.
[25] Arctic hares have up to 20% of their body mass as fat for insulation. This is a relatively low amount for a cold climate animal and the lack of weight that this brings enables them to reach 60km/h (40 mph) while running to escape predators. See Cool Antarctica Editors (2020), "Arctic Hare Facts and Adaptations *Lepus arcticus*", *Cool Antarctica*.
[26] Krzos (2006), p. 14.

mammoths were *not* polar creatures. French prehistorian Henry Neuville conducted the most detailed study of mammoth skin and hair. At the end of this thorough analysis, he wrote the following:

> It appears to me impossible to find, in the anatomical examination of the skin and [hair], **any argument in favor of adaptation to the cold.**[27]

Last, but not least, the mammoths' diet argues against the creature existing in a polar climate. How could they sustain their vegetarian diet of hundreds of pounds of daily food in an arctic region devoid of vegetation for most of the year? How could they find the gallons of water that they had to drink each day?

To make things worse, the woolly mammoth lived during the last ice age, when temperatures were markedly colder than today. Mammoths could not have survived the harsh northern Siberian climate of today, even less so 12,900 years ago when it should have been significantly colder.[28]

The evidence strongly suggests that the woolly mammoth was not a polar creature but a *temperate* one. Consequently, at the beginning of the Younger Dryas, 12,900 years ago, Siberia was not an arctic region but a temperate one.

---

[27] Henri Neuville (1919), "On the Extinction of the Mammoth," *Annual Report of the Smithsonian Institution*, p. 332.

[28] Isotope data suggest that before the onset of the Younger Dryas, temperatures were about 5°C lower than they are today. See next chapter for more detail.

The Younger Dryas

The Younger Dryas is named after a flower[29] that grows in cold conditions and became common in Europe during this cold period that started around 12,900 years ago and lasted for about 1,200 years. The Younger Dryas marks the transition between the Pleistocene epoch and our current epoch, known as the Holocene.

The Younger Dryas saw a sharp decline in temperature over most of the Northern Hemisphere.[30] It was the most recent and longest interruption to the gradual warming of the Earth's climate. To give an idea of the magnitude of the cooling, the Greenland ice core GISP2 indicates that, at its height, it was between 10°C[31] and 15°C[32] colder[33] during the Younger Dryas than it is today.

Figure 6: Younger Dryas flowers

Notice, however, that the overall cooling that occurred during the Younger Dryas was not homogeneous,[34] and while some regions like Siberia, Europe, Greenland or Alaska experienced a marked cooling, other regions like Northern America, apart from Alaska, and the 'Asian' side of Antarctica experienced a relative warming. Keep that in mind because this is an important point that we'll explore soon.[35]

Along with the drastic temperature drop, one of the major features of the Younger Dryas is a *massive die-off*: 35 mammals, among them mastodons, giant beavers, saber-toothed cats, giant sloths and woolly rhinoceroses, and 19 genera of birds, became extinct in a very short time period.[36]

It is estimated that as many as 40 million animals died in North America alone.[37] In total, hundreds of thousands of mammoths were killed. Mammoth remains have been found all across

---

[29] *Dryas octopetala*. The Younger Dryas, but also the Older Dryas and the Oldest Dryas, are named after *Dryas octopetala*, because of the great quantities of its pollen found in cores dating from those times.

[30] Don Easterbrook, (2019), "Younger Dryas – Climatology," *Encyclopaedia Britannica*.

[31] C. Buizert *et al.* (2014), "Greenland temperature response to climate forcing during the last deglaciation," *Science* 345 (6201): 1177–1180.

[32] Jeffrey Severinghaus *et al.* (1998), "Timing of abrupt climate change at the end of the Younger Dryas interval from thermally fractionated gases in polar ice," *Nature* 391 (6663):141–146.

[33] Between 18°F and 27°F.

[34] Malory Marcia (2013), "New evidence that cosmic impact caused Younger Dryas extinctions," *Phys.org*.

[35] See chapter "Crustal slippage."

[36] Katherine Bagley (2009), "Scientists Find Smoking Gun for California Comet Theory," *National Audubon Society*.

[37] Charles Hapgood (1999), *The Path of the Pole*, Adventure Unlimited Press, p. 250.

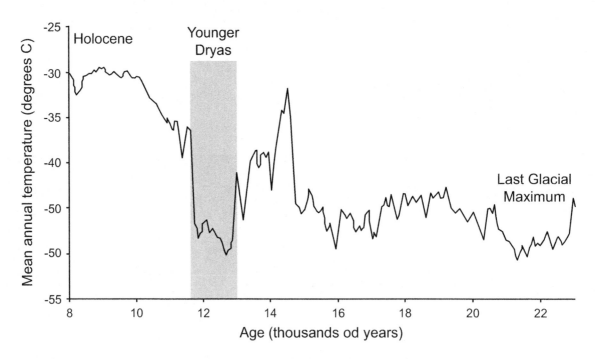

Figure 7: Mean annual temperature (22,000 to 8,000 BP)

Northern Russia from the Urals to the Bering Strait and on the American continent in Alaska and Yukon.[38] Only two small pockets of mammoths remained: Saint Paul Island in Alaska until 5,600 years ago and Wrangel Island in Siberia until 4,020 years ago.[39]

Human populations were already quite widespread at the time, including the Yurok, Hopies, Kato, Arawaks, Toltecs and Incas. We know for sure that at least one of these peoples, the Clovis[40] – who inhabited Northern America – was erased from the face of Earth during this period of turmoil. The Clovis people were not a small, localized tribe; their implantation sites cover most of Northern America, as indicated by the geographic range of their artifacts, particularly fluted points.

The map on the left below shows the widespread distribution of Clovis fluted points ca. 13,000 BP. The map on the right shows the disappearance of Clovis fluted points and their partial replacement by post-Clovis artifacts ca. 12,000 BP:

---

[38] M. Nuttall (2005), *Encyclopedia of the Arctic*, New York: Routledge Publishing.

[39] Russell Graham *et al.* (2016), "Holocene mammoth population extinction," *Proceedings of the National Academy of Sciences* 113(33):9310-9314.

[40] The Clovis culture is named for distinct stone tools found near Clovis, New Mexico, in the beginning of the 20th century. It appears 13,200 to 12,900 years ago and is characterized by the manufacture of "Clovis points" and distinctive bone and ivory tools. The Clovis culture ended ca. 12,900 BP and was replaced by several smaller and more localized regional societies. See: Mario Pino *et al.* (2019), "Sedimentary record from Patagonia, southern Chile supports cosmic-impact triggering of biomass burning, climate change, and megafaunal extinctions at 12.8 ka," *Scientific Reports* 9(4413):13.

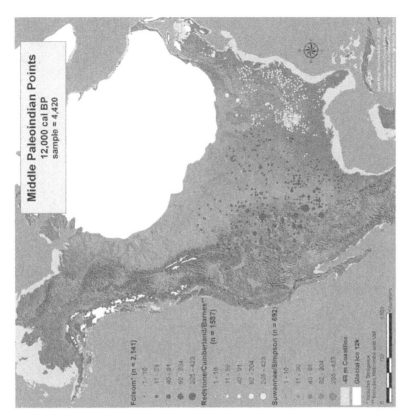
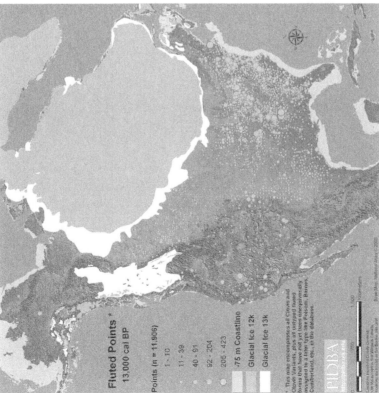

Figure 8: © PIDBA
Discovery sites of Clovis fluted points

The Crime Scene

The broad geographic scale of the extinction and its relatively recent occurrence provide a lot of scientific material. In the numerous excavations conducted over most parts of the northern hemisphere, the burial sites of the woolly mammoths reveal the very same features again and again:

Soot: a concentration peak in charcoal and soot[41] was found in several Clovis sites and in the Younger Dryas strata.

Figure 9: Charcoal and soot strata (also known as black mat) delineating the Clovis period

Fullerenes: a pure form of carbon like graphite and diamond. It is a large spheroidal molecule consisting of a hollow cage of sixty or more atoms of carbon. High concentrations of fullerenes[42] were found in the 12,900 BP strata.

Potassium 40: a naturally radioactive isotope with a 1.3-billion-year half-life that represents a tiny fraction of all potassium on Earth. This amount is very uniform throughout the solar system, *except for meteorites, comets, or when a supernova is involved*. Peak concentration of

---

[41] Adrienne Stich *et al.* (2012), "Soot as Evidence for Widespread Fires at the Younger Dryas Onset (YDB, 12.9 ka)," *Quaternary International* 279–280:468.

[42] L. Becker *et al.* (2009), "Wildfires, Soot and Fullerenes in the 12,900 ka Younger Dryas boundary layer in North America," *American Geophysical Union*.

this isotope[43] was found in the Clovis strata.

Helium-3: A typical extraterrestrial impact marker, Helium-3 is rare on Earth but common in extraterrestrial material. The connection between asteroid impacts and helium-3 was demonstrated by American astrobiologist Luann Becker,[44] who located a 40-kilometer-wide[45] impact site called the Bedout[46] crater that dates to the Permian extinction, 250 million years ago, and revealed high levels of helium-3. Similarly, the Younger Dryas boundary contains peak concentrations of helium-3.[47]

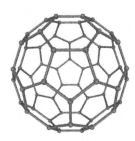

Figure 10: © CC BY-SA 3.0 Model of the C60 fullerene

Thorium, Titanium, Cobalt, Nickel, Uranium and other rare earth elements: High concentrations of these elements[48] were found in the Younger Dryas strata, Clovis sites and several meteorite craters. These rare elements are seldom found on Earth but they are very common in meteorites.

Carbon glass: the 12,900 BP strata is characterized by a high concentration of this form of black glass[49] that is rich in carbon. Testing showed that the carbon glass samples included numerous internal gas bubbles.[50] This is an indication of extraordinarily high temperatures followed by very sudden cooling; pure carbon melts at 3,500°C.[51] Only extraordinary events such as a major cometary impact can generate such temperatures. Carbon glass is only found in the Clovis-era layer.

Figure 11: © Firestone
Carbon glass found in three different Clovis sites

---

[43] Rodney Chilton (2009), *Sudden Cold: An Examination of the Younger Dryas Cold*, First Choice Books.

[44] Luann Becker *et al.* (2001), "Impact Event at the Permian-Triassic Boundary: Evidence from Extraterrestrial Noble Gases in Fullerenes," *Science* 291(1530).

[45] 25 miles.

[46] Bedout High is a geological and geophysical feature centered about 250 km (160 miles) off the northwestern coast of Australia. It is a roughly circular area about 30 km (19 miles) in diameter where older rocks have been uplifted as much as 4 km (2.5 miles) towards the surface. See: J.D. Gorter (1996), "Speculation on the origin of the Bedout high – a large, circular structure of pre-Mesozoic age in the offshore Canning Basin, Western Australia," *PESA News*, pp. 32-34.

[47] R. B. Firestone (2010), "Analysis of the Younger Dryas Impact Layer," *Lawrence Berkeley National Laboratory*.

[48] Robert J. Tuttle (2012), *The Fourth Source: Effects of Natural Nuclear Reactors*, Universal Publishers, p. 263.

[49] Richard B. Firestone *et al.* (2007), "Evidence for an extraterrestrial impact 12,900 years ago that contributed to the megafaunal extinctions and the Younger Dryas cooling," *Proceedings of the National Academy of Sciences* 104(41):16016-16021.

[50] Ted Bunch *et al.* (2012), "Melted glass from a cosmic impact 12,900 years ago," *Proceedings of the National Academy of Sciences* 109(28):E1903-E1912.

[51] 6,400°F.

Platinum: In 2013, Harvard researcher Michail Petaev[52] discovered an anomalously large platinum (Pt) peak in ice core samples from the Greenland Ice Sheet Project 2 (GISP2). Petaev provided some of the most compelling evidence of an extraterrestrial impact at the onset of the Younger Dryas by conducting high-resolution time series – about 3-year increments – of ice core samples. He noted a rise in Pt concentrations over 14 years and a subsequent drop during the following 7 years, consistent with the known residence time of stratospheric dust.[53] Since then, additional research has confirmed that the platinum spike ca. 12,900 BP affected not only Greenland but the whole planet.[54]

Iridium: an extremely rare element in the Earth's crust but typically found in meteorites and asteroid material.[55] Geological strata relating to notable cometary bombardments include the dinosaur extinction dated to 65 million years BP, commonly referred to as the Cretaceous-Tertiary (K-T) Extinction,[56] as well as the Triassic-Jurassic Extinction,[57] which occurred approximately 200 million years BP. Both extinction events exhibit abnormally high concentrations of iridium. So does the Younger Dryas strata[58] ca. 12,900 BP.[59]

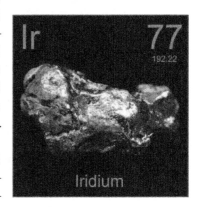

Figure 12: © periodictable.com Iridium sample

Nanodiamonds: Millions of microscopic diamonds were found in Clovis sites dating to ca. 12,900 BP. Hexagonal nanodiamonds[60] require pressures of 170,000 bars[61] and temperatures between 1,000–1,700°C[62] followed by rapid quenching in order to form. These are exceptional temperature and pressure conditions, but cometary impacts can easily create them.[63]

---

[52] Michail Petaev et al. (2013), "Large Pt anomaly in the Greenland ice core points to a cataclysm at the onset of Younger Dryas," *Proceedings of the National Academy of Sciences* 110:12917–12920.

[53] Ibid.

[54] Christopher Moore et al. (2017), "Widespread platinum anomaly documented at the Younger Dryas onset in North American sedimentary sequences," *Scientific Reports* 7(44031).

[55] L. Alvarez (1980), "Extraterrestrial Cause for the Cretaceous-Tertiary Extinction," *Science* 208(4448):1095–1108.

[56] Richard Cowen et al. (2000), *History of Life*, Blackwell Science, chapter "The K-T extinction."

[57] Tetsuji Onoue et al. (2012), "Deep-sea record of impact apparently unrelated to mass extinction in the Late Triassic," *Proceedings of the National Academy of Sciences* 109(47):19134-9.

[58] Rex Dalton (2007), "Archaeology: Blast in the past?" *Nature* 447(7142):256ff.

[59] While the Platinum spike is large and global, the iridium spike is not found in every ice core sample. An extraterrestrial impact rich in platinum but poor in iridium might come from a kind of meteorite called "magmatic iron." See: Robert Kunzig (2013), "Did a Comet Really Kill the Mammoths 12,900 Years Ago?" *National Geographic*.

[60] Douglas Kennett et al. (2009), "Shock-synthesized hexagonal diamonds in Younger Dryas boundary sediments," *Proceedings of the National Academy of Sciences* 106(31):12623–12628.

[61] 2 million psi.

[62] 1,800–3,100°F.

[63] R.W.K. Potter et al. (2013), "Numerical modeling of asteroid survivability and possible scenarios for the Morokweng crater-forming impact," *Meteoritics & Planetary Science* 48:744–757.

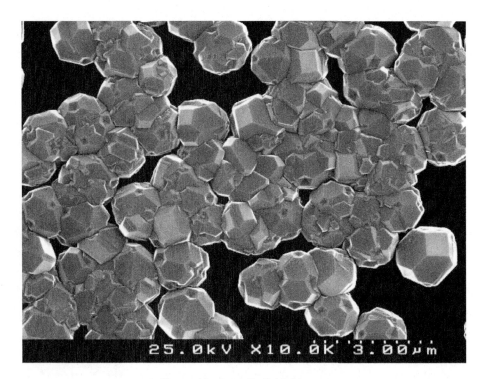

Figure 13: © 4.0 D. S Mukherjee
Synthetic nanodiamond crystals averaging about 100nm in size.

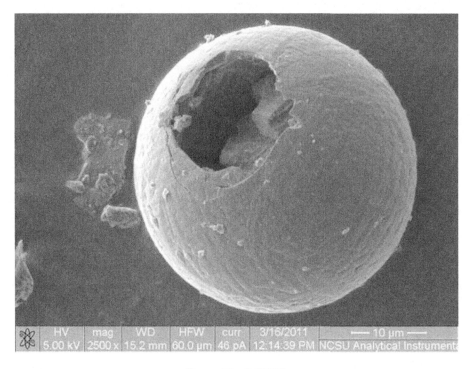

Figure 14: © PNAS
Magnetic microspherule found in the Younger Dryas boundary

Spherules: Hollow magnetic floating balls exhibiting a high concentration in carbon were found in most Clovis-era sites. These spheres are tiny, measuring from 10 to 50 micrometers[64] in diameter.[65] This form of carbon requires very high temperature and pressure to form but are very frequent in the Younger Dryas boundary, where thousands of spherules were found in each kilogram of dirt.

This long list of material – atypical isotopes like helium-3 and potassium-40, and rare elements like platinum, iridium, thorium and uranium – reveal again and again the same pattern. They're almost absent from our natural environment but *common in comets*, and are found in high concentration in Clovis-era strata and along impact craters.

Exotic materials like carbon glass, spherules, microscopic diamonds and fullerenes tell a similar story. They indicate exceptionally high temperatures and pressures that don't occur on Earth *except during extreme events like asteroid impacts*. All this material has been found in high concentration in impact sites and in the Clovis strata.

Firestone summarizes the results of years of research conducted in numerous geological sites across Europe and America in the following terms:

> In stratified sections at each of 10 [Younger Dryas] sites, from California to Belgium and Manitoba to Arizona, we found a <5 cm[66] thick sediment layer dated ca. 12.9 ka that **contained a majority of 14 markers**, forming distinct stratigraphic peaks at above background concentrations.
>
> These markers include magnetic microspherules (up to 2144/kg), magnetic grains (16g/kg) enriched in iridium (117 ppb, 6000 × terrestrial values), vesicular carbon spherules (1458 units/kg), glass-like carbon (16 g/kg), nanodiamonds, fullerenes containing extraterrestrial concentrations of 3He (84 × air), and soot and charcoal (2 g/kg).
>
> Except for small quantities of magnetic grains and charcoal, the **markers were undetectable in the sediment either above or below the impact layer**, representing stratigraphic sequences spanning >55 k.y. This is inconsistent with Pinter and Ishman's assertion of a **"constant" rain of meteoritic debris and demonstrates that a layer of concentrated extraterrestrial (ET) markers was deposited ca. 12.9 ka.**[67]

Along with the Younger Dryas boundary, there is a second boundary that contains similar high concentrations of extraterrestrial impact material: the K-T boundary mentioned above, also known as the Cretaceous-Paleogene transition, associated with the notorious Chicxulub Impact,[68] which marked the mass extinction that wiped out the dinosaurs.

The numerous discoveries of cometary material and impact material in the Clovis strata, in the K/T strata and in cometary/meteoric craters strongly suggest that *a massive cometary bombardment occurred about 12,900 years ago.*

---

[64] 0.4 to 2 thousandths of an inch.
[65] Firestone *et al.* (2007).
[66] 2 inches.
[67] Richard Firestone (2008), "Impacts, mega-tsunami, and other extraordinary claims," *GSA Today*.
[68] Cowen (2000), chapter "The K-T extinction."

The 'Event'

If the Younger Dryas and its accompanying mass extinction were caused by cometary bombardment, the next step is to identify the characteristics of the offending extraterrestrial bodies: their nature, size, angle of impact and, of course, location of impact.

In his book, *The Cycle of Cosmic Catastrophes*,[69] Firestone did an exceptional job gathering evidence about the asteroid impacts that triggered the beginning of the Younger Dryas. This book is a must-read if you want to know more details on this topic.

The first task was to identify where the cometary fragments hit Earth. To do so Firestone investigated "secondary craters," i.e. the craters created by ejecta coming from the primary impacts. Interestingly, the orientation of the secondary craters all pointed towards the same location.

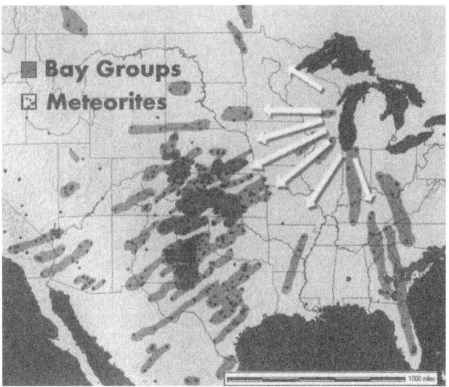

Figure 15: © Firestone
Orientation of the secondary impacts related to the body that hit Lake Michigan

---

[69]Richard Firestone *et al.* (2006), *The Cycle of Cosmic Catastrophes: How a Stone-Age Comet Changed the Course of World Culture*, Simon and Schuster.

As shown by the picture above, Firestone triangulated the trajectories followed by the ejecta[70] and identified five potential primary impacts as well as their estimated diameter:

- *Hudson Bay*, Canada: 480 km[71] diameter

- *Amundsen Bay*, Canada: 241 km[72] diameter

- *Baffin Island*, Canada: 120 km[73] diameter

- *Lake Saimaa*, Finland: 290 km[74] diameter

- *Lake Michigan*, United States: 105 km[75] diameter as shown in the picture below:

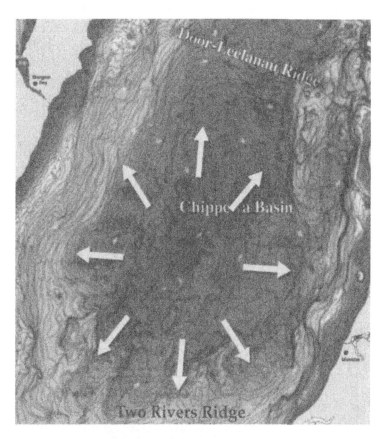

Figure 16: © Firestone
A potential impact location: the Chippewa Basin in Lake Michigan

---

[70] J.Y. Kaminski *et al.* (2004), "A General Framework for Trajectory Triangulation," *Journal of Mathematical Imaging and Vision* 21:27–41.
[71] 300 miles.
[72] 150 miles.
[73] 75 miles.
[74] 180 miles.
[75] 65 miles.

The next step was to check if there was any trace of primary craters in those five locations. And indeed there were. The primary craters were, however, noticeably shallower than expected. The shallowness of the craters relative to their width and length suggests that the impactors were not solid rock (meteorites) but more likely 'dirty snowballs' (cometary material) and that their impact angle was low.[76]

Indeed, the angle and the nature of the bolide have a direct influence on the shape of the crater.[77] A rock-solid meteorite following a vertical trajectory would create a round and deep crater, while a 'fluffy' cometary fragment hitting Earth at a low angle would create *a shallow elongated (elliptical) crater*.

Firestone's hypothesis was confirmed by geological surveys. For example, in Lake Michigan it was revealed that the Chippewa basin looked like a typical crater sub-basin, with its "terrace faulting,"[78] a stair-step pattern formed when large slabs of rock crack and slide downward after impact as depicted in the drawing below:

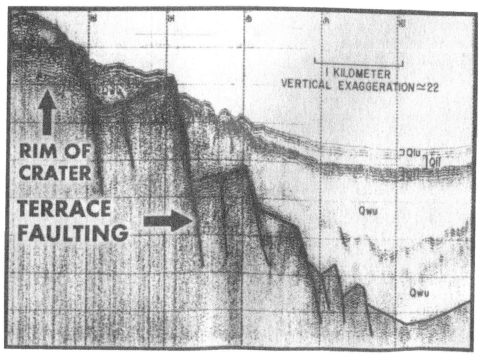

Figure 17: © Firestone
Seismic profile of the Chippewa Basin showing terrace faulting

The Chippewa basin also reveals some radial fracturing,[79] which is usually associated with extraterrestrial impacts.

---

[76] Firestone *et al.* (2006), chapter 19.
[77] V. Shuvalov (2009), "Atmospheric erosion induced by oblique impacts," *Meteoritics & Planetary Science* 44(8):1095–1105.
[78] Firestone *et al.* (2006), chapter 26.
[79] Fractures found along the radius of impact craters. See: Kumar Senthil *et al.* (2008), "Impact fracturing and structural modification of sedimentary rocks at Meteor Crater, Arizona," *Journal Of Geophysical Research* 113:E09009.

In summary, Firestone's bombardment scenario goes as follow: a massive comet approached Earth and fragmented into bolides of various sizes. Five cometary fragments were particularly massive and reached the surface of the planet.

The five impacts happened *virtually at the same time*, which suggest they were all part of the same cometary swarm. The fact that the first impacts in the list above hit a very specific location (Northern America) suggests that the comet fragmented into four pieces shortly before impact. The fifth crater located in Finland suggests that, before the final fragmentation, an earlier fragmentation had occurred, separating the body that hit Finland from the rest of the cometary body. Analysis of the five primary craters showed that they had a similar orientation; therefore they probably came from the same place and belonged to the same cometary cluster as indicated by the picture below.

Firestone could only check crater sites on the ground or under relatively shallow water. It's entirely possible that additional massive cometary fragments hit the oceans, particularly in the submerged vicinity of the five listed impacts (Arctic Ocean, northern Atlantic Ocean, Baltic Sea, etc.). Such impacts would leave no trace but would be nonetheless very destructive, creating massive tidal swells, among other effects.

Firestone also managed to estimate the angle of the cometary fragments by analyzing the geometry of the craters. The rings at the bottom of the craters have the same elliptical shapes. To create such elongated elliptical craters, the impactors would have had to come in at a low angle, between 5 and 15 degrees above the horizon.

Now we know that the woolly mammoths were killed by cometary bombardments. But the main question remains unanswered: how did the mammoths get flash-frozen? First, let's define more precisely what flash-freezing is.

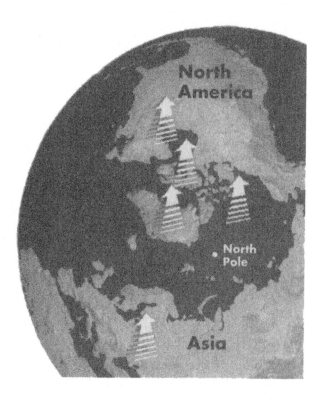

Figure 18: © Firestone
Similar direction exhibited by the 5 cometary fragments that hit Earth ca. 12,900 BP

What Is Flash-Freezing?

Flash-freezing is the sudden exposure of an item (like food or a biological sample) to cold temperatures in order to preserve it. American inventor Clarence Birdseye[81] developed the quick-freezing process of food preservation in the 20th century.

This rapid freezing is usually done by submerging the sample in liquid nitrogen[82] or a mixture of dry ice[83] and ethanol. Liquids are preferred over gases for cooling application because, typically liquids exhibits a thermal conductivity that is about 40 times greater than air.[84]

Figure 19: © Comet Corn
Comet corn,[80] an example of liquid nitrogen flash freezing

There are many forms of flash-freezing, from the mildest to the most sudden. So, what kind of flash-freezing was experienced by the woolly mammoths? Here is Mark Krzos's answer[85]:

> At normal body temperatures, stomach acids and enzymes break down vegetable material within an hour. What inhibited this process? The only plausible explanation is for the stomach to cool to about 40°F[86] in ten hours or less. But because the stomach is protected inside a warm body (96.6°F[87] for elephants), how cold must the outside air become to drop the stomach's temperature to 40°F? Experiments have shown that the **outer layers of skin would have had to drop suddenly to at least -175°F**[88]!

According to Russian scientist Sukachev[89] and other researchers[90] the undigested food like

---

[81] American inventor, entrepreneur, and naturalist (1886-1956). Considered to be the founder of the modern frozen food industry. See: Wikipedia Editors (2004), "Clarence Birdseye," *Wikipedia*.

[81] Comet Corn is a Californian brand of flash frozen organic popcorn. The mention of the word comet is purely fortuitous and doesn't refer to cometary-induced flash-freezing; it is an interesting coincidence nonetheless.

[82] Nitrogen is in liquid form at -195°C (-320°F).

[83] "Dry ice" is the solid form of carbon dioxide. It sublimates (phase transition from solid to gaseous) at -78°C (-109°F).

[84] Q.T. Pham (2014), *Food Freezing and Thawing Calculations*, Springer.

[85] Krzos (2006), p. 49.

[86] 4°C.

[87] 36°C.

[88] -115°C.

[89] V.N. Sukachev (1914), "The study of plant remains from the food mammoth found on the Berezovka River," Nauchnye rezul'taty ekspeditsii, *snaryazhennoi Akademiyei Nauk dlya raskopki mamonta*, naidennogo po r. Berezovke 1901(3):1e17.

[90] See for example: Frederick G. Wright (1902), *Asiatic Russia*, McClure, Phillips and Co., or J. Felix (1912), "Das

grasses, mosses, shrubs and tree leaves found in the stomachs and digestive tracts of the mammoths is not the only evidence of flash-freezing.

Food was also found in the *mouths* of the frozen mammoths. This food, consisting mainly of buttercups,[91] had been cropped but not chewed or swallowed. The buttercups froze so rapidly that *they still had the imprint of the mammoths' molars*. Despite their elasticity, these buttercups did not have time to revert back to their initial shape after the mammoth died.

For biological applications,[92] the key idea of flash-freezing is to drop the temperature fast enough *so that large ice crystals cannot form* and burst or puncture cells. This is exactly what was revealed to be the result of a detailed analysis of the cellular samples extracted from the woolly mammoths:

> The flesh of many of the animals found in the muck must have been very **rapidly and deeply frozen, for its cells [had] not burst.** Frozen-food experts have pointed out that to do this, starting with a healthy, live specimen, you would have to drop the temperature of the air surrounding it to a point well below minus 150 degrees[93]Fahrenheit.[94]

Figure 20: © AFP/Getty
The frozen mammoth found in Lyakhovsky

The picture above shows a female mammoth in pristine condition, found on the Lyakhovsky Islands[95] in 2013. Unexpectedly, when scientists poked at the mammoth's frozen remains with

---

Mammuth von Boma," *Veröffentlichungen des Städtischen Museum for Völkerkunde zu Leipzig* 4:1–55.

[91] John Morris (1993), "Did the Frozen Mammoths Die in the Flood or in the Ice Age?" *Institute For Creation Research.*

[92] University of Florida discussion board (2012), "Freezing Tissue," *Biotech.ufl.edu.*

[93] -101°C.

[94] Ivan Sanderson (1960), "Riddle of the Frozen Giants," *Saturday Evening Post*, p. 82.

[95] Southernmost group of the Siberian Islands in the arctic seas of eastern Russia.

an ice pick, *blood started to flow.*[96] This suggests that the woolly mammoths froze faster than the few minutes it would have taken for their blood to coagulate.

According to experts, for this sudden freezing to occur inside the warm body of woolly mammoths who had a thick layer of fat, they would had to have been subjected to extremely low temperatures: -115 degrees Celsius.[97]

If we posit that temperatures in Siberia, which was under a temperate climate at the time, were around 16°C,[98] it means that the temperature dropped from +16°C to –115°C, *a 131 degrees Celsius*[99] *temperature drop within a few hours at most.* This is a huge number, and we have to wonder, has such a severe drop in temperature ever occurred on our planet?

---

[96] Colin Schultz (2013), "Scientists Just Found a Woolly Mammoth That Still Had Liquid Blood," *Smithsonian Magazine*.
[97] -175°F. It is the same temperature mentioned by Skraz in order to inhibit digestion within an hour.
[98] 60°F.
[99] 235°F.

Recorded Cases of Sudden Cooling on Our Planet

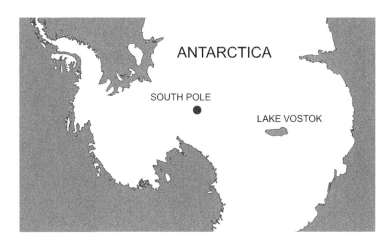

Figure 21: Location of the coldest place on Earth: Lake Vostok, Antarctica

The largest temperature drop ever recorded happened on September 11, 2011, over the U.S. Midwest when an exceptional thunderstorm[100] caused a drop of 39°C (from 27°C to -12°C[101]) in 14 hours.[102] Yet this record pales in comparison to what happened to the mammoths both in terms of geographic scope and temperature magnitude.

The current record low temperature is -89°C,[103] measured at Vostok. It still falls short of the requirements to flash-freeze a mammoth. In addition, Vostok is located close to the center of Antarctica, which experiences 6-month-long winter nights and subsequent freezing temperatures. This place is much colder than the temperate regions that the mammoths inhabited.

Notice that while -150°F is required to flash-freeze a mammoth, a higher temperature could achieve the same result *if sufficient wind was generated*. This phenomenon is called 'wind chill,' as shown in the wind-chill table below. For example, an air temperature of -76°F[104] combined with 55mph[105] winds leads to a heat loss equivalent to -150°F, i.e. the temperature required to flash-freeze mammoths and other animals. While a -76°F temperature cannot be experienced at

---

[100] This storm was nicknamed "the Great Blue Norther." Its cold front brought at least nine tornadoes (including one F4 tornado) which touched down across five states, resulting in 13 fatalities.
[101] From 80°F to 11°F. That is a 69°F drop.
[102] Patrick Market (2011), "A Record Setting Temperature Shift 100 Years Ago Today," *CBS Saint Louis*.
[103] Exactly −89.2°C or −128.6°F measured on ground level at the Soviet Vostok Station in Antarctica on July 21, 1983. See: J. Turner *et al.* (2009), "Record low surface air temperature at Vostok station, Antarctica," *Journal of Geophysical Research* 114:D24.
[104] -60°C.
[105] 90 km/h.

ground level in temperate regions, *this is a fairly common occurrence not far above our heads*:

As indicated in the below diagram, 90 km[106] above our heads the average atmospheric temperature is -90°C.[107]

At an altitude of 11 km,[108] the average temperature is about -60C.[109] The 11 km altitude marks the area where the upper limit of the troposphere and the lower limit of the stratosphere meet. This boundary is called the 'tropopause.'

The problem is that *the tropopause forms an almost unbreakable boundary.* Only three types of documented extreme events can trigger a breach of the tropopause:

- superderechos,[110] which are a kind of super-storm,

- pyrocumulonimbus,[111] which are giant smoke clouds triggered by massive fires, and

- major volcanic eruptions.[112]

Figure 22: © Meteo France
2015 Etna eruption displaying a mushroom-shaped column reaching the top of the troposphere.

However, such events are by definition local, and therefore can't account for the flash-freezing of the whole Siberian region along with part of Alaska and the Yukon.

So what could have carried the freezing upper atmosphere air down onto a vast region of the planet? Could cometary bodies be the real culprits?

It may seem counter-intuitive that the impact of an asteroid could cause massive cooling on the surface of our planet – after all, when entering the atmosphere, rocks heat up and, on reaching the surface of the planet, will spread fire and heat. That *is* true, but as we will now see this is only part of the story.

---

[106]60 miles.
[107]-130°F.
[108]7 miles.
[109]-76°F.
[110]Stephen Corfidi *et al.* (2013), "About Derechos," Storm Prediction Center, *NOAA Web Site*.
[111]Michael Fromm *et al.* (2000), "Observations of boreal forest fire smoke in the stratosphere by POAM III, SAGE II, and lidar in 1998," *Geophysical Research Letters* 27(9):1407–1410.
[112]T. Mather, D. Pyle, and C. Oppenheimer (2013), "Tropospheric Volcanic Aerosol," in *Volcanism and the Earth's Atmosphere* (Ed. A. Robock and C. Oppenheimer).

Temperature (°F)

Wind (mph)	Calm	40	35	30	25	20	15	10	5	0	-5	-10	-15	-20	-25	-30	-35	-40	-45
5		36	31	25	19	13	7	1	-5	-11	-16	-22	-28	-34	-40	-46	-52	-57	-63
10		34	27	21	15	9	3	-4	-10	-16	-22	-28	-35	-41	-47	-53	-59	-66	-72
15		32	25	19	13	6	0	-7	-13	-19	-26	-32	-39	-45	-51	-58	-64	-71	-77
20		30	24	17	11	4	-2	-9	-15	-22	-29	-35	-42	-48	-55	-61	-68	-74	-81
25		29	23	16	9	3	-4	-11	-17	-24	-31	-37	-44	-51	-58	-64	-71	-78	-84
30		28	22	15	8	1	-5	-12	-19	-26	-33	-39	-46	-53	-60	-67	-73	-80	-87
35		28	21	14	7	0	-7	-14	-21	-27	-34	-41	-48	-55	-62	-69	-76	-82	-89
40		27	20	13	6	-1	-8	-15	-22	-29	-36	-43	-50	-57	-64	-71	-78	-84	-91
45		26	19	12	5	-2	-9	-16	-23	-30	-37	-44	-51	-58	-65	-72	-79	-86	-93
50		26	19	12	4	-3	-10	-17	-24	-31	-38	-45	-52	-60	-67	-74	-81	-88	-95
55		25	18	11	4	-3	-11	-18	-25	-32	-39	-46	-54	-61	-68	-75	-82	-89	-97
60		25	17	10	3	-4	-11	-19	-26	-33	-40	-48	-55	-62	-69	-76	-84	-91	-98

Figure 23: © NOAA
Wind chill table

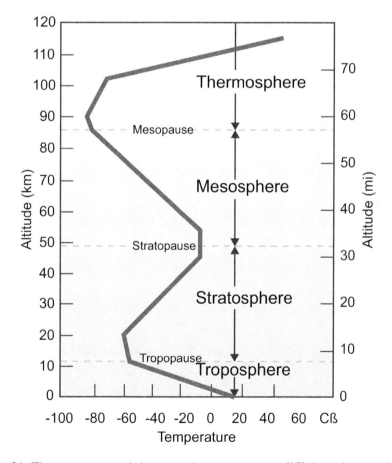

Figure 24: The tropopause and the atmospheric temperature (°C) depending on elevation

Atmospheric Ablation Induced by a Cometary Impact

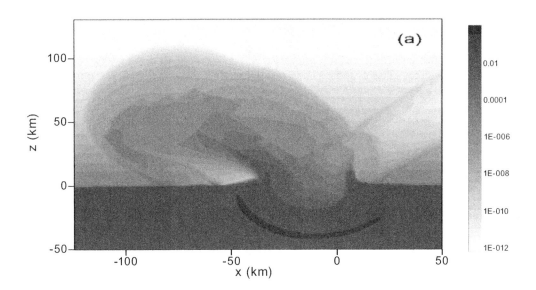

Figure 25: © Shuvalov
Analysis of an oblique asteroid impact. Density distributions are shown. The plume expands outside the wake in an oblique impact.

Until recently, asteroids were solely considered as the harbingers of fire and scorching heat. However, in 1983 scientists came up with the concept of asteroid-induced atmospheric erosion.[113]

When large and fast enough, an asteroid can *obliterate part of the Earth's atmosphere.* On impact, the asteroid vaporizes (heat and pressure transform the asteroid into gases), as does a similar mass of the Earth's surface where the impact occurs.

The ensuing hot gas plume can expand *faster than escape velocity*,[114] which is about 11.2 km/s[115] on Earth.[116] For comparison, the typical speed of asteroids in space[117] is about 30 km/s.[118] The escaping plume drives off the overlying air into space.

---

[113] Shuvalov (2009).

[114] Escape velocity is the minimum speed needed for a free, non-propelled object to escape from the gravitational influence of a massive body, that is, to achieve an infinite distance from it. Escape velocity is a function of the mass of the body and distance to the center of mass of the body. See: Wikipedia Editors (2002), "Escape Velocity," *Wikipedia*.

[115] 25,000 mph.

[116] Shelley Canright (2009), "Escape Velocity: Fun and Games," *NASA website*.

[117] Anurag Ghosh (2010), "What is the Average Asteroid Speed?" *BrightHub website*.

[118] 67,000 mph.

Specifically, the part of the atmosphere that will be carried up into space along with the hot gas plume is shaped like an inverted cone. It is known as the 'cone' of atmospheric erosion.[119]

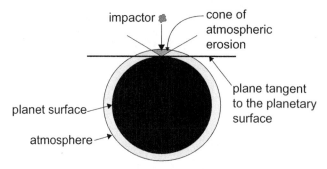

Fig. 7: In impact erosion of a planetary atmosphere, greater energy of impact leads to a wider cone of atmospheric ablation (orange zone) until the entire atmosphere above a plane tangent to the planet is removed.

Figure 26: © Catling
Impact erosion of the Earth's atmosphere

The shape of this cone will depend on the size of the asteroid, its density, speed and angle relative to the Earth's surface.

To better understand atmospheric erosion, let's look at an analogous phenomenon that we are familiar with, called 'water backdrop.'

When you drop an object in water, you'll sometimes witness water moving upwards from the location were the object impacted the water. The water acts like a spring and recoils upwards. This recoil can take the form of a water plume and/or water drops.

In a similar way after the impact of an asteroid, matter and gases will move upwards because of the recoil effect, plus they will be boosted by the rising ambient heat. But, unlike the water drop, they won't fall back down, because the speed of the rising material exceeds escape velocity – the velocity required to escape the gravity of our planet.

The drawing below is inspired by the work of Valery Shuvalov,[120] who calculated the effects of cometary/asteroid bombardment in terms of atmospheric erosion. The cases Shuvalov studied are, however, limited to bodies that are smaller and exhibit a greater impact angle than the cometary fragments hypothesized to have hit Hudson Bay 12,900 years ago. I've tried to apply Shuvalov's analysis to the Hudson object posited by Firestone:

Figure 27: Water backdrop on top of a water plume

---

[119] D. C. Catling, and K. J. Zahnle (2009), "The escape of planetary atmospheres," *Scientific American* 300:36–43.
[120] Russian volcanologist at the Institute of Geospheres Dynamics, member of Russian Academy of Sciences.

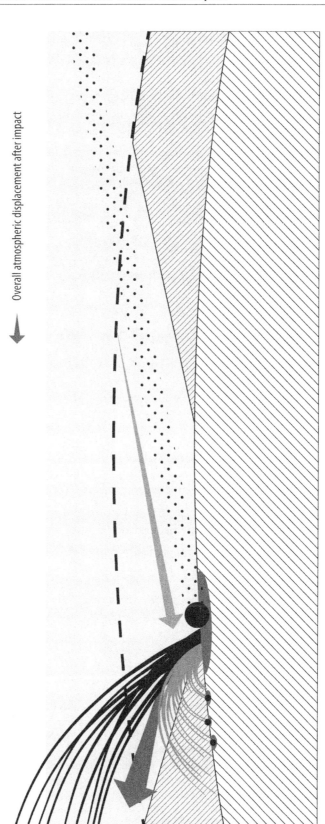

Figure 28: Impact of a 50-mile diameter cometary fragment at 15° angle

The cometary fragment (black ball) is estimated to be 80 km[121] in diameter and entered the atmosphere from the north at a low angle – about 15° – as depicted by the dotted wake.

At impact, the cometary fragment created a substantial but shallow primary crater about 200 km[122] in diameter, as shown by the dark grey crater, and a massive plume, which is represented by the medium grey and black ejecta in the drawing, which will, in turn, create debris-induced secondary craters as exemplified by the Carolina Bays,[123] which are depicted in the following photograph:

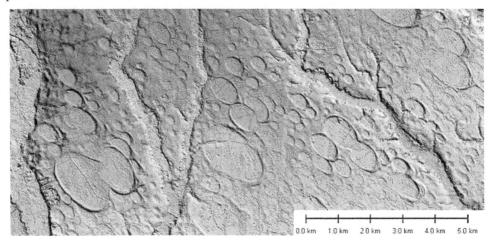

Figure 29: © LIDAR Flood Mapping Project
LIDAR[124] image of Carolina Bays in Robeson County, N.C.

Notice also the cone of atmospheric ablation, the light grey area below the dotted line, which represents the upper boundary of the atmosphere before ablation. The diameter of the cone at ground level is about 1,000 km.[125] The non-ablated part of the atmosphere is the thin dashed area visible on the left and on the right of the drawing.

Of course, one single drawing doesn't convey the magnitude of the forces and dynamics that are likely present during such an impact, so allow me to explain further:

- At first the atmosphere around the cometary body is accelerated through friction as illustrated by the medium grey arrow along the dotted wake, similar to the wind you feel when standing by a car that passes nearby, albeit much more powerful.

- At impact, the powerful wind created along the wake combines with the massive stream of

---

[121] 50 miles.

[122] 300 miles.

[123] The Carolina Bays are shallow, elliptical craters concentrated along the U.S. Atlantic seaboard and covering nine states from Delaware to Florida. There are possibly 500,000 Carolina Bay–type features (up to 10 km long) on Earth, probably many more. See: W.F. Prouty (1952), "Carolina Bays and their Origin," *Geological Society of America Bulletin* 63(2):167–224.

[124] LIDAR is a method for measuring distances by illuminating the target with laser light and measuring the reflection with a sensor. LIDAR is a combination of the terms "LIght" and "raDAR"

See: Jie Shan *et al.* (2018), *Topographic Laser Ranging and Scanning: Principles and Processing*, second edition, *CRC Press*.

[125] 700 miles.

super-hot gases and vaporized material, part of which reaches escape velocity and flies out into space in a massive updraft as shown by the dark grey arrow in the drawing, *carrying with it a large chunk of Earth's atmosphere* represented by the ejecta in black. Meanwhile, the slowest ejecta fall back to Earth's surface as shown by the medium grey ejecta.

- For a short period after the impact, *the ablation zone is a void of empty space* as represented by the medium grey area. For reference, the temperature of outer space is -270.5°C,[126] while the temperature of space near the Earth is about 10°C.[127]

- The vacuum is followed by a downdraft equally powerful to the updraft that preceded it. *Super-cooled air violently refills the void.*

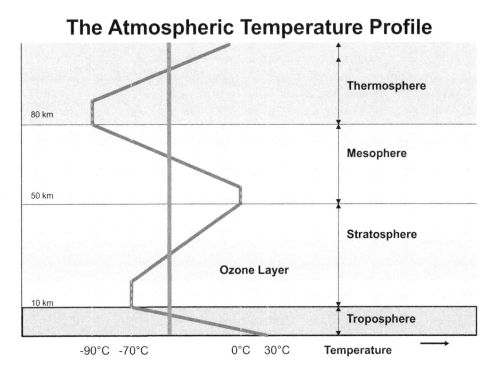

Figure 30: Temperature of the atmospheric layers

This downdraft consists of air mostly located in the various layers of the upper atmosphere. Because higher atmosphere is less dense, its molecules move faster.

In the high atmosphere, as indicated by the thick vertical line in the diagram above, the temperature is about -50°C[128] on average, but atmospheric temperatures can be as low as -90°C[129] just above the mesopause.

---

[126]-455°F. Angela Libal (2018), "The Temperatures of Outer Space Around the Earth," *Sciencing.com*.
[127]50°F.
[128]-58°F.
[129]-130°F.

The whole refilling process involves super-cooled[130] air, because the surrounding air when filling the vacuum will encounter a pressure drop.

In addition, with part of the atmosphere having been ablated, the atmosphere as a whole loses volume and ends up being thinner, which leads to an overall atmospheric pressure drop (decrease in the height of the atmospheric column).

Depression cools gases: you have an example when you use a compressed air duster to clean your keyboard: while the pressure in the can drops, the air and the can get colder and colder.

We have described these three atmospheric features:

- tornado-speed winds,

- cold air inflow from the upper atmosphere,

- super-cooling due to decompression,

which may actually combine to generate truly unfathomable chill factors that could have easily flash-frozen the woolly mammoths and numerous other animals.

Now that we have an idea of how the woolly mammoths and friends got flash-frozen, the next question is, *how did they remain frozen for thirteen millennia?*

To do so, they would have had to remain in an environment where temperatures stayed below 0°C.[131] Aside from the ice sheets, such conditions on Earth occur only in permafrost,[132] which is found in mountains and/or at high latitude (60° or more).

But northern Siberia doesn't feature any high mountains and its latitude, at the time, was about 40° north, as shown by the temperate flora and fauna found in this region at the time. This means that Siberia experienced well *above* freezing temperatures most of the year.

To explain how the mammoths remained frozen for about 13,000 years, we must introduce the notion of wandering geographic poles.

---

[130] In this context super-cooling refers to the process of lowering the temperature of the atmosphere below its freezing point without it becoming a solid. Our atmosphere is mostly nitrogen whose freezing point is -210°C (-346°F), but nitrogen can remain gaseous at lower temperature if its pressure is low enough.

[131] -32°F.

[132] Heather Doyle (2020), "What Is Permafrost?" *NASA Climate Kids*.

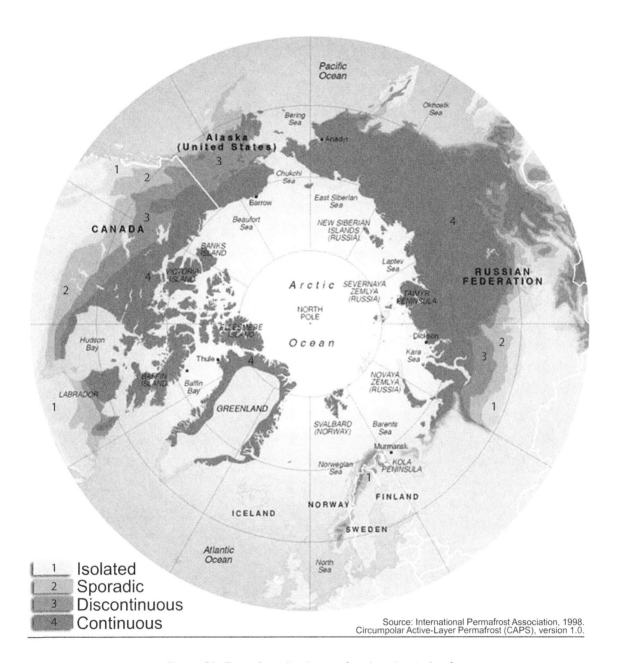

Figure 31: Permafrost distribution (northern hemisphere)

Wandering Geographic Poles

It is commonly believed that the geographical poles[133] have always been in the same present location. However, data proves that this is not the case. Actually, the location of the poles has changed a lot, even in recent times.

Some of the best evidence that the geographic poles have changed location is coral. Coral reefs require a minimum water temperature of about 22°C.[134] However geologic analysis reveals the presence of coral in some of today's coldest areas:

> In the Carboniferous[135] formation we again meet with plant remains and beds of true coal in the Arctic regions. Lepidodendrons and calamites, together with large spreading ferns, are **found at Spitzbergen, and at Bear Island in the extreme north of Eastern Siberia;** while marine deposits of the same age contain an abundance of large stony corals.[136]

For decades, Chinese oceanographer Ting Ying Ma[137] studied coral and managed to establish the positions of the ancient coral-lines, which more or less coincide with the equator line in ancient times as shown in the map below.

Ma's coral-lines/equator line run in all directions – **one even crosses the Arctic Ocean**. Ancient coral colonies have also been found in Ellesmere Island,[138] within the Arctic Circle.

Another way to determine the past location of the geographic poles is called paleomagnetism, which is the analysis of the direction of iron particles in rocks like magnetite or hematite.

When these rocks form while solidifying from a liquid phase, in the context of a volcanic eruption for example, the magnetized iron particles in the molten rocks act as compasses and solidify in a position that lines up with Earth's magnetic field.

Not only do these iron particles indicate the direction of the magnetic north at some time in the past, but because of their vertical dip they also indicate how far away the pole is, i.e. the latitude. The closer to the pole the iron particle is, the less vertically tilted it will be.

One problem with this method is the fact that the magnetic pole also wanders. However, over a period of a few thousand years the magnetic pole returns to its original position, and the average position of the magnetic pole over the whole period coincides with the Earth's axis of rotation.

Therefore, for paleomagnetism to reliably reveal the position of the geographic pole, samples over a long enough period of time have to be collected. That's the reason why lava flows are so

---

[133] The geographic poles are the axis around which the earth spins. They are different from the magnetic poles, which are determined by the direction a compass points.
[134] 72°F. Sam Kahng *et al.* (2012), "Ecology of mesophotic coral reefs – Temperature related depth limits of warm-water corals," *Proceedings of the 12th International Coral Reef Symposium*.
[135] The Carboniferous is a geologic period that spans from the end of the Devonian Period 360 Mya to the beginning of the Permian Period 300 Mya.
[136] Hapgood (1999).
[137] Catherine Shu (2011), "A shift in time," *Taipei Times*.
[138] Robert Bast (2009), "Crustal Poleshifts," *Survive2012*.

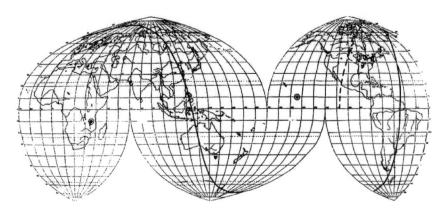

Figure 32: © Ting Ying Ma
The coral line during the Silurian epoch (about 430 Mya)

valuable. Eruption after eruption, they stack over each other while each lava flow indicates the location of the pole at the time on an eruption.

American professor Charles Hapgood compiled the locations of the geographic poles over the ages and found a higher than expected variability. During the Pleistocene,[139] the geographic pole occupied *23 different locations*. From the end of the Precambrian epoch[140] until now, Hapgood identified a grand total of *229 different locations* for the geographic pole,[141] as shown in the table below:

Epoch or Period	Duration (Millions of years)	Eur.	Asia	India	N. Amer.	S. Amer.	Afr.	Aust.	Antarc.
Pleistocene	1	6	15	0	1	1	?	?	?
Tertiary	62	0	8	1	4	0	0	1	1
Cretaceous	73	0	3	10	1	0	1	1	0
Jurassic	46	6	2	1	0	1	7	1	3
Triassic	49	8	7	0	20	0	2	1	0
Permian	50	15	0	0	9	0	2	2	1
Carboniferous	65	15	1	0	5	0	4	2	0
Devonian	50	7	0	0	0	0	0	2	2
Silurian	20	0	2	0	2	0	1	3	0
Ordovician	75	0	2	0	1	0	1	3	0
Cambrian	100	1	0	0	3	0	0	2	0
Precambrian	?	3	0	0	22	0	0	3	0
		61	40	12	68	2	18	21	7
							Grand Total of Poles..		229

Figure 33: © Hapgood
Locations of the geographic pole since the Precambrian

---

[139] The Pleistocene is the epoch that lasted from about 2.5 Mya and ended with the beginning of the Younger Dryas ca. 12,900 years ago.
[140] The Precambrian is the period that spans from the time all rocks were formed to the beginning of the Cambrian period about 500 million years ago.
[141] Hapgood (1970), p. 12.

Now that we know that the location of the geographic poles is not as fixed as once believed, let's try to determine where the geographic pole was located before the impacts in question.

Location of Geographic North Pole before Impact

Geology provides a robust method to determine the past location of ice caps and as a result, the past location of the geographic poles – the pole being roughly at the center of the ice cap.

The edges of the ice cap are set in motion by the pressure of the ice behind them, and their scouring action striates the bedrock of the continent over which the ice cap expands.[142]

Figure 34: © ramalama_22
Glacier-induced striation in Central Park (New York)

Geological studies[143] indicate that before the Younger Dryas, during the last phase of the Pleistocene (from 17,000 BP to 13,000 BP), the northern ice sheet, called the Laurentide[144] ice sheet, was centered over Hudson Bay.

---

[142]Don Easterbrook (1999), *Surface Processes and Landforms*, Prentice Hall.
[143]W.F. Ruddiman *et al.* (1981), "The mode and mechanism of the last deglaciation: Oceanic evidence," *Quaternary Research*.
[144]Laurentide is a region of Québec, Canada.

The Laurentide ice sheet covered almost all of Canada, Greenland (except its coast) and a small fraction of Northern Europe. All the rest of the Northern hemisphere, including the Arctic Ocean, Alaska, Siberia and part of the Yukon *were ice free*, as shown by the map below:

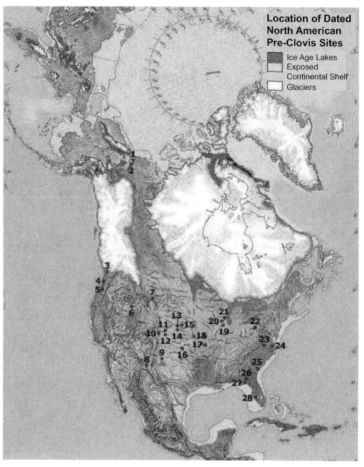

Figure 35: © NOVA
The Northern ice sheet (ca. 13,000 BP)

As noticed by Hapgood, the Laurentide ice sheet was similar both in size and shape to the present Arctic ice sheet:

> The first line of evidence that the last North American ice cap was a polar ice cap is based on the shape, size, and peculiar geographical location of the ice sheet. Two geologists, Kelly and Dachille, have pointed out that **the area occupied by the ice was similar both in shape and size to the present Arctic Circle.**
> Many others have remarked on its unnatural location. It seems to have occupied the northeastern rather than the northern half of the continent. **No one has explained why the ice cap, which extended southward as far as Ohio, did not cover some of the northern islands of the Canadian Arctic Archipelago, islands lying between Hudson Bay and the pole or why it failed to cover the Yukon District of Canada or the northern part of Greenland.**[145]

---

[145]Hapgood (1970), p. 216.

The above strongly suggests that before the Younger Dryas, the north geographic pole was located in the vicinity of Hudson Bay, which is about 60° north or *30 degrees in latitude away from the current north pole.*

The peculiar Laurentide ice sheet is not the only evidence available. The study of fossils provides a very good idea of what kind of plants and animals lived in different locations of the planet right before the Younger Dryas. This research tends to confirm that, at the end of the Pleistocene, the north pole was located in Hudson Bay.

Indeed, before the Younger Dryas, the Arctic Ocean was a temperate ocean as indicated by the presence of warm-water[146] and foramanifera[147] in sea cores.[148] Siberia was a temperate region, as shown by its temperate fauna and flora.[149] Japan was warmer than today, as indicated by flora that grows in a temperate climate and by the corals of Okinawa.[150]

Another piece of evidence comes from Antarctica. A geographic north pole located around Hudson Bay would place the geographic south pole – as indicated by the black dot in the map below – about seven times farther away from the Ross Sea[151] – circled in black – than it is now:

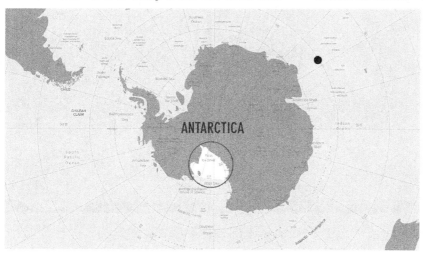

Figure 36: © Oceanwide Expeditions
The Ross sea is circled. The black dot indicates the antipode of Hudson Bay.

If the Ross Sea was so much farther from the pole at the end of the Pleistocene ca. 13,000 BP, it should not have been glaciated, and this is exactly what sea cores reveal, with layers of fine sediment typical of temperate climates.[152] This fine sediment is carried by rivers from *ice-free*

---

[146] N.S. Oskina *et al.* (2019), "Warm-Water Planktonic Foraminifera in Kara Sea Sediments," *Oceanology* 59:440–450.
[147] Foramaniferas are amoeba-like sea creatures. They secrete a tiny shell about 1 mm long.
[148] NASA Editors (2005), "A Record from the Deep: Fossil Chemistry," *NASA Earth Observatory.*
[149] Krzos (2006), p. 15.
[150] Arisa Seki *et al.* (2012), "Mid-Holocene sea-surface temperature reconstruction using fossil corals from Kume Island, Ryukyu, Japan," *Geochemical Journal* 46(3).
[151] Ross Sea is a bay of the Southern Ocean in Antarctica. It is named after British explorer James Ross who visited the area in 1841. The Ross Sea is only 320 km (200 miles) from the current geographic south pole.
[152] Brenda Hall (2013), "History of the grounded ice sheet in the Ross Sea sector of Antarctica during the Last Glacial Maximum and the last termination," *Geological Society,* Special Publications 381.

continents. The past ice-free nature of Antarctica is corroborated by ancient maps depicting the continent devoid of any ice. The topic of ancient maps showing Antarctica is extensively covered later in this book,[153] so we will leave that for now.

Interestingly, if the north pole was located in Hudson Bay prior to the Younger Dryas, it would explain a mystery that has puzzled many experts.

Some ancient constructions like Stonehenge and Teotihuacan[154] display an awkward orientation. The main axes of these two sites point roughly towards the north, but not exactly. Teotihuacan is 15° off while Stonehenge is about 40° off.

However, as depicted in the map below *both constructions point directly towards Hudson Bay*. One might wonder if Stonehenge and Teotihuacan were built prior to the Younger Dryas and aligned with the north-south axis of the time.

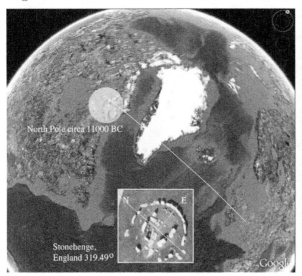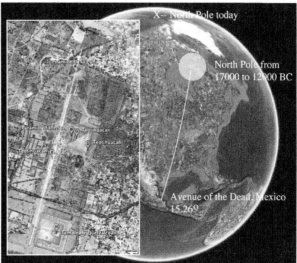

Figure 37: © Viewzone.com
Alignment of the Teotihuacan and Stonehenge with Hudson Bay

The location of the Laurentide ice sheet, the fauna and flora fossils found in Arctic, the sediments and maps indicating an ice-free Antarctica, and the alignment of very ancient constructions all suggest that, about 13,000 years ago the geographic north pole was located around Hudson Bay, which is roughly 60°N, or 30° away from the current north pole.

Consequently, the geographic south pole was not located in Antarctica, *and Northern Siberia was at 40° degrees of latitude north*[155], *which is the current latitude of Spain, Greece, Italy or California*. This latitude is typical of temperate climate. It is under this temperate latitude that the woolly mammoths lived, but it's *not* under this latitude that their corpses were preserved in ice.

The cometary bombardment had dramatic effects for our planet, including a modification of the location of its geographic poles. Now, let's look at how this might have happened.

---

[153] Part 2, chapter "When did the water transfer occur?"
[154] Dan Eden, "Changing Poles" (2011), Chapter 2, *Viewzone*.
[155] The current longitude of Northern Siberia is 70°N, from which we subtract 30°, giving 40°N.

Crustal Slippage

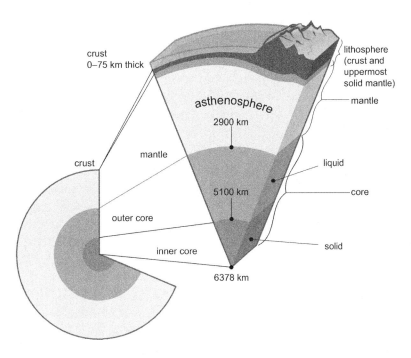

Figure 38: © Jeremy Kemp
Earth's internal structure (not to scale)

We might consider our planet to be a solid piece of rock because all we see is its surface made of solid rock, like mountains, deserts or ocean floors. However, as depicted by the diagram above, solid rock is only a tiny part of planet Earth, a thin layer called 'crust' that is less than 100 km[156] thick.

Beyond the crust is the mantle, a thick layer of magma with an average thickness of about 3,000 kilometers.[157] While the lower part of the mantle behaves like a solid because of the tremendous pressures that bind the molten material together, the upper mantle, also known as the asthenosphere, which is both hot and under relatively little pressure, exhibits a low viscosity[158] and *has the mechanical properties of a semi-fluid*.

These kinds of fluid properties are particularly present in a specific layer of the asthenosphere:

> It seems that such a layer has been discovered in the asthenosphere at a depth of about 100 miles.[159] According to the Soviet geophysicist V. V. Beloussov, chemical processes at this depth,

---

[156] 60 miles.
[157] 1,800 miles. Fraser Cain (2016), "What is the Earth's Mantle Made Of?" *Universe Today*.
[158] Kent Condie (2016), *Earth as an Evolving Planetary System*, Academic Press, Chapter 4.

made possible by change of phase, are changing heavier into lighter rock, thus causing gravitational instability as the lighter rock tries to rise to the surface. Beloussov has named this the "wave-guide layer". Observations by the American geophysicist Frank Press are in general agreement. Press finds (from satellite observations) that **this layer is a very liquid one. It seems that if the earth's outer shell does slide as a unit over the interior, this is the most likely level at which the movement can occur.**[160]

So, from a mechanical perspective, the crust is relatively similar to an iceberg floating on the ocean (the low-viscosity, liquid asthenosphere). The low viscosity of the upper mantle explains why continents keep on drifting. It also suggests that *it takes far less mechanical force to shift the crust relative to the mantle than to shift the whole planet.*

Shifting the whole planet – crust, mantle and core – would require enormous forces. When one enters the data relative to the asteroids posited by Firestone (about 80 km[161] in diameter) in an asteroid impact simulator,[162] the energy of Firestone's cometary fragment is much too low to have induced a change in the orbit of our planet, its tilt or even its spin rate. Such an object could not budge our planet, which, comparatively, carries far too much momentum.

For comparison, the estimated mass of Earth is about 6×10^{24} kg,[163] while the mass of an 80 km[164] diameter asteroid is about 6.10^{14} kg. The Earth is therefore about 10 billion times heavier than the cometary fragment.

However, the low viscosity of the upper mantle might have allowed such impacts to cause the crust to slip relative to the mantle, particularly if the incoming asteroids exhibited a low angle of entry, which seems to be the case for the cometary fragments posited by Firestone. The next picture shows part of the physics modeling of an asteroid-induced crustal slippage. I'll spare you the math beyond the phenomenon, which has been fully theorized by B. L. Freeborn.[165]

According to Charles Hapgood, the cometary bombardment that triggered the Younger Dryas caused the Earth's crust to slip by about 30°, moving the geographic poles to their current location. For Italian engineer Flavio Barbiero,[166] the crust slipped by about 20° degrees.

Hapgood and Barbiero might be close to the truth. In any case, the slippage must have been *greater* than 20° in order to shift Siberia into the permafrost region,[167] which encompasses latitudes greater than 60°N, and would allow the mammoths to remain frozen.

This crustal slippage might explain why the temperature change that occurred during the Younger Dryas was not homogeneous, as previously mentioned.[168] Regions like Siberia, Europe,

---

[159] 160 km.

[160] Hapgood (1970), p. 119.

[161] 50 miles.

[162] See for example the impact calculator developed by ESA and the Faulkes Telescope based on the calculations of Robert Marcus *et al.* http://simulator.down2earth.eu/planet.html?lang=en-US

[163] Brian Luzum *et al.* (2011), "The IAU 2009 system of astronomical constants: The report of the IAU working group on numerical standards for Fundamental Astronomy," *Celestial Mechanics and Dynamical Astronomy* 110(4):293–304.

[164] 50 miles.

[165] B. L. Freeborn (2014), *The Deep Mystery: The Day the Pole Moved*, Tiw & Elddir Books.

[166] Flavio Barbiero (2020), "Changes in the Rotation Axis of Earth After Asteroid/Cometary Impacts and their Geological Effects," *International Journal of Anthropology* 35(1-2).

[167] Karen Frey (2009), "Impacts of Permafrost Degradation on Arctic River Biogeochemistry," *Hydrological Processes* 23:169–182.

[168] See chapter "Younger Dryas."

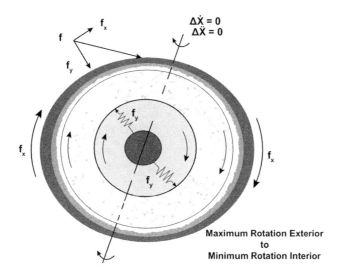

Figure 39: © B.L. Freeborn
Mechanical simulation of an asteroid-induced crustal slippage.

Greenland or Alaska shifted closer to the new geographic north pole and consequently experienced a marked cooling; other regions like North America (apart from Alaska) and the 'Asian' side of the Antarctic moved away from the new geographic north pole and experienced a relative warming.

At this point we have a good idea of how the mammoths may have been flash-frozen and why they remained frozen for millennia. However, the corpses of the mammoths revealed several other puzzling pieces of evidence.

The Coroner's Verdict

Since 1800, at least 11 scientific expeditions,[169] mostly in Northern Siberia, have excavated frozen mammoths and other mammals, including ox, wolverine, voles, squirrels, bison, rabbit and lynx.[170]

Of the numerous frozen mammoth finds, the Berezovka mammoth[171] is probably the most famous. It was found frozen along the Berezovka River[172] – hence its name – in a near-perfect state of preservation. Only part of the trunk and head had to be reconstructed because they were not embedded in ice and were consequently eaten by predators.

The Berezovka mammoth (image below) is displayed in the Zoological Museum in St. Petersburg, Russia, in the struggling position in which it was found near the Berezovka River, just inside the Arctic Circle.

The pristine state of the frozen mammoths allowed scientists to extract a lot of information about the mammoths and their cause of death. In fact, the mammoths are so well preserved that some scientists[173] are attempting to splice mammoth DNA with Asian elephants to reintroduce the extinct species.

Coroners who examined many mammoths *found several recurring features*:

Fractures: the Berezovka mammoth (and others) had numerous broken bones,[174] including several ribs, shoulder blade and pelvis.

Dirt and suffocation: dirt was found in the lungs and digestive tracks of frozen mammoths.[175] The only cause of death that could be definitely established was *suffocation*. No other cause of death has been established for the remaining woolly mammoths.

Figure 40: © JohntheFinn
The Berezovka stuffed mammoth as displayed in the St. Petersburg Museum

---

[169]Harris Allen (2011), *Somewhere in the Bible: Understanding Bible Scriptures and Creation*, WestBow Press, p. 215.

[170]Paul LaViolette (2005), *Earth Under Fire: Humanity's Survival of the Ice Age*, Bear & Co., p. 211.

[171]Hans Krause (1978), *The Mammoth—In Ice and Snow?*, Stuttgart, Germany, self-published, "Chapter 1: The Berezovka Mammoth."

[172]Short river located in the Altaï Republic in Western Siberia.

[173]Sarah Pruitt (2017), "Scientists Say They Could Bring Back Woolly Mammoths Within Two Years," *History.com*.

[174]J.T. Hagstrum *et al.* (2017), "Impact-related microspherules in Late Pleistocene Alaskan and Yukon 'muck' deposits signify recurrent episodes of catastrophic emplacement," *Scientific Reports* 7(16620).

[175]Krzos (2006), p. 79.

The second buried mammoth found by Russian researcher Vollosovitch[176] had an erect penis, and its cause of death was also deemed to be suffocation. A Woolly mammoth named Dima was found to have pulmonary edema,[177] suggesting death by asphyxia after great exertion just before death.

Yedomas: these are mounds, 10 to 90 meters tall,[178] made of *soil mixed with thick veins of ice*. Yedomas are widespread in Siberia,[179] where the total area of its occurrence is about 1 million km^2.

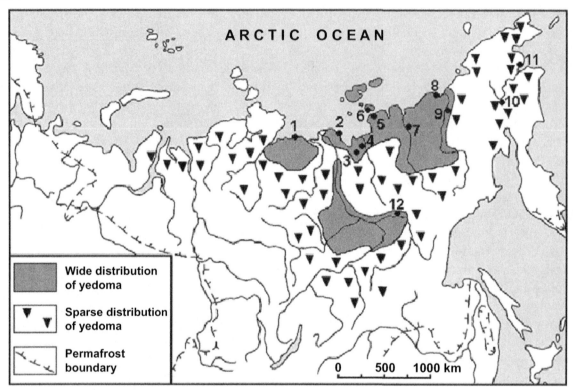

Figure 41: © Mikhail Kanevskiy
Yedoma geographic distribution in Russia

Yedomas are very *rich in carbon* and are literally *filled with dead trees and dead animals*.[180] For example, the 'mammoth cemetery' is a yedoma containing the corpses of not less than 156 mammoths.[181] The soil constituting the yedomas is called 'loess,'[182] which is basically *wind-blown silt*, in other terms eolian sediment.

---

[176] R. Dale Guthrie (1990), *Frozen Fauna of the Mammoth Steppe*, The University of Chicago Press.
[177] Errol Fuller (2004), *Mammoths: Giants of the Ice Age*, Bunker Hill Publishing, Inc.
[178] 30 to 260 feet.
[179] Mikhail Kanevski *et al.* (2011), "Cryostratigraphy of late Pleistocene syngenetic permafrost (Yedoma) in northern Alaska, Itkillik River exposure," *Quaternary Research* 75(3):584–596.
[180] Sergey Zimov *et al.* (2014), "Role of Megafauna and Frozen Soil in the Atmospheric CH4 Dynamics," *Public Library of Science*.
[181] John Massey Stewart (1977), "Frozen Mammoths from Siberia Bring the Ice Ages to Vivid Life," *Smithsonian*.
[182] M. Frechen (2011), "Loess in Europe," *Quaternary Science Journal* 60(1).

Figure 42: Cross section of a yedoma exposed by river erosion

<u>Upright position</u>: a number of mammoths, including the one at Berezovka, were found frozen in an upright position.[183]

Now that we know the *features* of the cometary bombardment and the *evidence* found by the mammoths' coroners, we can attempt to combine them and reconstruct the catastrophic timeline that sealed the tragic fate of these creatures.

Each location experienced its own variation on the disaster theme. Describing each combination of effects experienced in each part of the globe would be too tedious and, after all, our main topic is the woolly mammoths. So in the following chapter we will focus on the sequence of events that occurred in Siberia that led to the demise of the woolly mammoths.

---

[183] Michael Oard (1979), "A Rapid Post-Flood Ice Age," *Creation Research Society Quarterly* 16:29–37.

The Tragic Fate of the Woolly Mammoths

As indicated by the discovery of the ripe fruits of sedge, grasses, and other plants in the stomach of the Berezovka mammoth, the action takes place around *summertime*,[184] *in a temperate and lush northern Siberia meadow*,[185] *12,900 years ago.*

First, a new star appeared in the night sky, and then it grew in magnitude. It became visible in daytime and eventually it outshone the sun, both in brightness and size.

A few minutes before impact, the 'second Sun' separated into at least 5 majors parts and many minor ones which crossed the sky above Siberia and followed a trajectory towards the north before disappearing beyond the horizon while traveling at about 35 km/s[186] or about 80,000 miles per hour.

Figure 43: Chelyabinsk meteor flash on February 15, 2013

The sky was streaked with the fiery wakes of thousands of small fragments that disintegrated in the atmosphere. A sudden wind triggered by the wake of the major cometary fragments started to blow up dirt from the ground and shake the surrounding trees.

Powered by the air depressions[187] created by the wake, the wind grew in force, the air became filled with dust, the mammoths braced themselves against the wind as they raised their heads and opened their mouths, gasping for some dust-free air.

The impact itself lit up the northern horizon, and the prolonged flash of light was blinding. The updraft plume threw a massive chunk of heated atmosphere into space. The pressure dropped suddenly, bringing down the atmospheric temperature and, for a few seconds, the mammoths were exposed to the glacial void of space, and the flash-freezing began. Some mammoths died at this point from suffocation.

This freezing in a vacuum might explain a very peculiar 'oxygen-free' ice that was found underneath the frozen mammoths:

> Deeper down in the cliff the ice becomes more solid and transparent, in some places entirely white

---

[184]Valentina Ukraintseva (2013), *Mammoths and the Environment*, Cambridge University Press, p. 237.
[185]Hapgood (1970), p. 268.
[186]20 miles per second.
[187]Shuvalov (2009).

and brittle. **After remaining exposed to the air even for a short time this ice again assumes a yellowish-brown color and then looks like the old ice.**

Obviously, something in the air (probably oxygen) reacted chemically with something in the ice. **Why was air (primarily oxygen and nitrogen) not already dissolved in the ice?** Just as liquid water dissolves table salt, sugar, and many other solids, water also dissolves gases in contact with it. For example, virtually all water and ice on Earth are nearly saturated with air. **Had air been dissolved in Herz's rock ice before it suddenly turned yellowish-brown, the chemical reaction would have already occurred.**[188]

Then hurricane force winds started blowing super-cooled air toward the impact zone to refill the vacuum. This supernatural, freezing wind lasted for hours. If the cone of ablation was 400 km[189] in radius, hurricane-force winds would blow for about two hours to refill the void. Such a wind would have frozen the mammoths, and many other creatures, to the core.

While some mammoths were frozen in an upright position, others were blown away and/or bombarded with flying debris (trees, boulders). This might explain the numerous broken bones found during the autopsies.

Along with this freezing wave, Siberia was inundated with *unprecedented rainfall*. The two main ingredients for precipitation are cooling and dust. The cooling leads to condensation – atmospheric water vapor transforming into liquid water – while the atmospheric dust acts as nucleation agents around which rain droplets form.

The magnitude of the cooling and the quantity of atmospheric dirt-saturated air led to torrential rainfall. In Siberia, where the cooling was the most marked, the deluge of hail and snow was the most severe.

Because of the quantity of dust, soot, dirt and sediment in the atmosphere, the falling hail and snow were very dirty and dumped exceptional quantities of frozen water, soot and sediment on Siberia.

Normally, these sheets of dirty rain, hail and snow would have stopped after a few days once atmospheric dust and water vapor were removed from the atmosphere by precipitation, but they continued because atmospheric dust and water vapor were *continually supplied by the ongoing ground and underwater volcanic eruptions*[190] triggered by the impact and the crustal slippage.

On top of this, the cooling was sustained and even worsened by an 'albedo loop,' where an increasing part of the planet's surface was covered with snow and ice that reflected more and more of the little sunlight that managed to penetrate the dusty atmosphere, which led to further cooling that created more ice and snow.

The diagram below describes the albedo loop and how sustained volcanic activity powered and worsened the whole albedo dynamic:

---

[188] Krzos (2006), p. 62.
[189] 250 miles.
[190] For an explanation about cometary impacts can trigger volcanic eruptions, see part III, chapter "Correlation between Cometary Activity and Volcanic Activity."

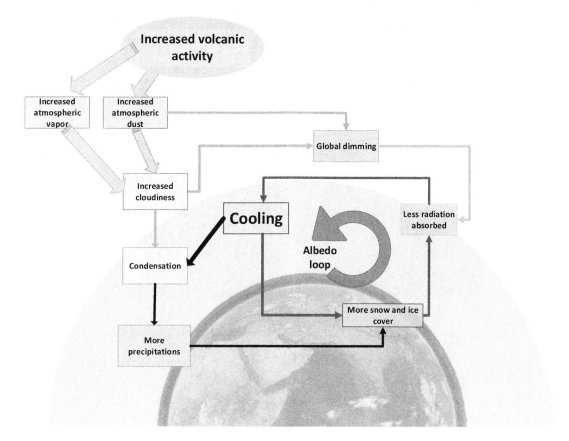

Figure 44: © Sott.net
Albedo loop and volcanic activity

In the above we have used mainstream concepts (dust, rain condensation, cooling) to explain the weather effects of the cometary bombardments. However, electricity also plays a major role, particularly when atmospheric dust is involved.

The role played by electricity in weather phenomena has been thoroughly covered in my previous book.[191] Below is a quick summary of the influence of electric charges and atmospheric dust on precipitation.

In fair weather conditions, electrons on the surface of the Earth are attracted by the positively charged ionosphere. If there is dust in the atmosphere, the free circulation of electrons is hindered and electrons are captured by the atmospheric dust, creating negatively charged areas in the atmosphere.

Those local atmospheric electric charges are what ultimately power hurricanes with their associated precipitation and lightning, which, ultimately, are charge-balancing phenomena that bring electrons back to the Earth's surface. In addition, electric charges catalyze the accretion of water droplets.[192]

---

[191] Pierre Lescaudron, with Laura Knight-Jadczyk (2014), *Earth Changes and the Human-Cosmic Connection*, Red Pill Press.

[192] G. Goyer (1960), "Effects of electric fields on water-droplet coalescence," *Journal of Meteorology* 17.

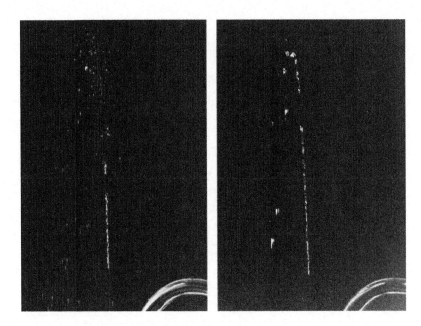

Figure 45: © Bounds
Influence of electric field on the size of water drops. On the right, droplets are subjected to an electrical field

The cometary impacts generated massive sources and amounts of atmospheric dust: impact craters, secondary craters, volcanic eruptions, giant forest fires triggered by the plume and the ejecta, and tornado-speed winds. In addition, the atmospheric dust generated by the cometary fragments was electrically charged.[193]

The electrically charged cometary fragments themselves disrupted the atmospheric electrical field between the Earth's surface and the ionosphere, which is the driver of precipitation and chaotic weather.[194]

The greatest recorded twenty-four-hour rainfall exceeded 180 cm.[195] It occurred under rather normal circumstances, so we can start to imagine the kind of precipitation that can be induced by the unique combination of catastrophic factors: dust-saturated atmosphere, electrically charged dust and a disrupted atmospheric electrical field.

Precipitation in the aftermath of the cometary impact carried tons of atmospheric dust down to Earth and might well be the cause for the yedomas, which are basically an accumulation of wind-blown sediment and frozen water.

This layer of wind-blown sediment must have covered many parts of the northern hemisphere. But, as shown by the dark grey areas in the map below, nowadays yedomas are only found in parts of Siberia and Alaska. This is because those regions are covered by permafrost that held the ice complex/yedoma together and prevented water erosion – rain and rivers – from washing it away into the oceans.

---

[193]Charles Bruce, "Comets," *Plasma Universe*.
[194]Lescaudron (2014), Chapter 26.
[195]72 inches. The event occurred in January 1966 during the passage of Tropical Cyclone Denise next to La Reunion Island, which is located east of Madagascar in the Indian Ocean. G.J. Holland (1993), "WMO/TC-No. 560, Report No. TCP-31," *World Meteorological Organization*, Geneva, Switzerland.

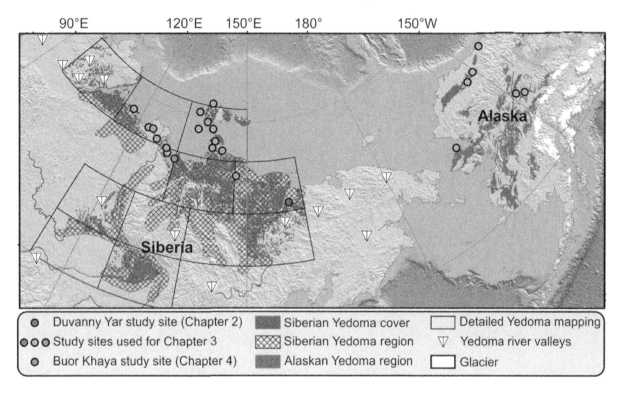

Figure 46: © Strauss
Yedoma geographic distribution

It was a truly apocalyptic scenario that is difficult to fathom. Perhaps the closest comparison would be a months-long giant glacial hurricane generating incredible winds, mountains of dirty hail and snow, flying trees and animals, rolling boulders with never-ending volcanic eruptions and earthquakes in the background.

Mega Tsunamis

Tsunamis were, of course, also part of the picture. Firestone's five cometary fragments impacted either oceans or ice sheets; each impact could, therefore, have triggered mega-tsunamis. The previously mentioned Hudson impactor, the largest body identified by Firestone,[196] directly hit the 3 kilometer-thick[197] ice cap and projected thousands of cubic kilometers[198] of melt-water onto the Atlantic Ocean.

At the time, however, the sea level was about 80 meters lower than it is today. So on-ground evidence for tsunamis is scarce but nonetheless existent in the form of giant traces of coastal flooding also known as 'chevron dunes' because of their typical shape:

Figure 47: © Google Earth
Chevron dunes of Ampalaza Bay (Madagascar)

---

[196] 480 km in diameter (300 mi).
[197] 2 miles.
[198] Alan Condron *et al.* (2012), "Meltwater routing and the Younger Dryas," *Proceedings of the National Academy of Sciences* 109(49):19928–19933.

Typically, a tsunami generated by a magnitude 7.5–8.0 quake reaches a height of between 3 and 10 meters.[199] The most powerful recorded earthquake occurred in Chile in May 1960 with a magnitude of 9.5,[200] but it "only" triggered a 15-meter-high tsunami,[201] which remains well below the 80-meter sea level rise mentioned earlier. For comparison, the Madagascar chevron dunes depicted in the satellite picture above reach more than 200 m above present sea level,[202] with inland penetration of up to 45 km.[203]

While 75% of tsunamis are due to earthquakes, up to 10% of all reported tsunamis, particularly the larger ones, are due to an "unidentified" cause. Cometary impact on oceanic areas might be the "unidentified" cause of those mega-tsunamis as demonstrated by the presence of impact craters and ejecta containing cometary material near the location of several mega-tsunamis.[204]

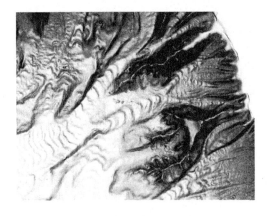

Figure 48: © Moore *et al.*
Sediment wave channels in southeastern Lake Tahoe

Several places reveal the marks of mega tsunamis generated by the cometary impacts ca. 12,900 BP. One of them is Lake Tahoe in the U.S., where the 'Great Lake Tahoe Comet Tsunami' occurred, creating a wave up to 300 meters high.[205] Another affected site is Lommel in Holland, where:

> A very interesting and unusual structure of the disturbance can be observed in a peat layer of the Usselo Horizon (Fig. 19). **This structure was interpreted as a possible fingerprint of a paleotsunami. Providing the age of the Usselo Horizon [12,900 BP], such a tsunami might be related to the impact in question.**[206]

The Gulf of Cadiz in Spain tells a similar story:

> The results of sedimentological analyses, ichnological treatment, spatial distributions, and radiocarbon dating of the YSLs [Younger Dryas Sand Layers] suggest that the possibility of an origin of the YSLs is **tsunami-related**[207]

---

[199] 10 to 30 feet.
[200] K. Satake *et al.* (2007), "Long-Term Perspectives on Giant Earthquakes and Tsunamis at Subduction Zones," *Annual Review of Earth and Planetary Sciences* 35(1):355.
[201] 50 feet. D. Abbott *et al.* (2007), p. 3.
[202] 650 feet. D. Abbott *et al.* (2007), p. 10.
[203] 28 miles.
[204] D. Abbott *et al.* (2007), pp. 16–31.
[205] 1,000 feet. George Howard (2020), "Lake Tahoe Comet Tsunami," *The Cosmic Tusk*.
[206] George Howard (2011), "West to East?: Belgian Tsunami at the Lower Younger Dryas Boundary," *The Cosmic Tusk*.
[207] Yasuhiro Takashimizu *et al.* (2016), "Reworked tsunami deposits by bottom currents: Circumstantial evidences from Late Pleistocene to Early Holocene in the Gulf of Cádiz," *Marine Geology* 377:95–109.

No Human Remains But Archaeological Evidence

For decades researchers have been puzzled by the absence of human remains alongside the thousands of dead woolly mammoths and other Pleistocene mammals:

> No **human remains (even bones or teeth)**, no weapons (arrows or knives), and no other artifacts (pottery, utensils, or art) have been found alongside frozen mammoth and rhinoceros remains.[208]

The absence of human remains seems to be due to two factors.

First, before the Younger Dryas, the human population was very scarce, with an estimated ten million people on the planet,[209] which represents an average population density[210] of one human per 50 km^2.[211]

Second, more than 80% of the current world population lives within 800 km[212] of the oceans.[213] But, before the Younger Dryas, the sea level was about 80 meters lower than it is today, as shown in the diagram to the right.

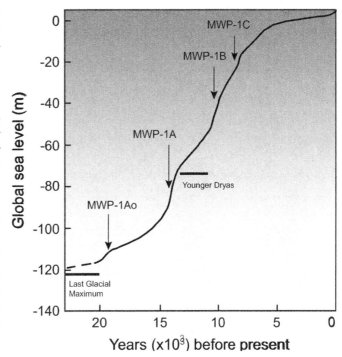

Figure 49: © NASA GISS
Sea level vs. global temperature (20,000 BP–Now)

---

[208] Krzos (2006), p. 196.
[209] Alfred J. Andrea (2011), *World History Encyclopedia, Volume 10*, ABC-CLIO, p. 160.
[210] For comparison, the estimated pre–Younger Dryas population represents a mere 0.12% of the current 7.5 billion world population. The pre–Younger Dryas world population was around 10,000,000 and Earth's total area 510,000,000 km^2 (197,000,000 sq. mi.), therefore the worldwide human population density was approximately 10,000,000 ÷ 510,000,000 = 0.02 per km^2 (0.06 per sq. mi.).
[211] 19 square miles.
[212] 500 mile.
[213] Asad Latif (2009), *Three Sides in Search of a Triangle: Singapore-America-India Relations*, Institute of Southeast Asian Studies, p. 46.

This means that, if like today, the pre–Younger Dryas human population were concentrated along the coastline, then *most previously inhabited pre–Younger Dryas sites are now fully covered by oceans.*

This point is especially true for the northern coast of Russia,[214] where the water depth of the continental shelf averages only 60 m[215] and extends more than 650 km[216] under the Arctic Ocean,[217] as depicted in the map below where the dotted black line shows the boundary (continental break) of the continental shelf:

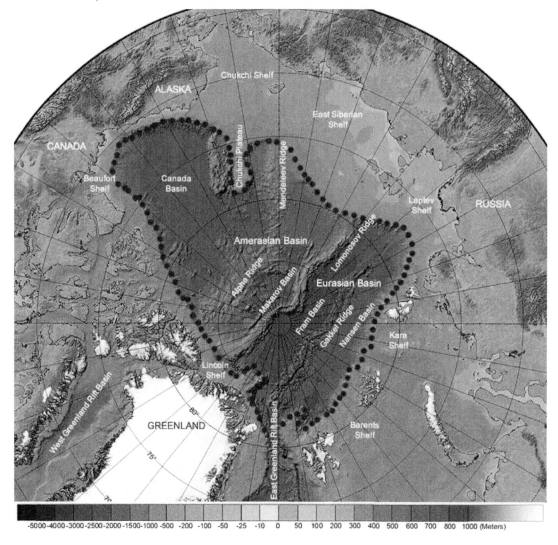

Figure 50: © IBCAO
Depth map of the Arctic Ocean

---

[214] Vladimir Isachenkov, "Russia to UN: We are claiming 463,000 square miles of the Arctic" (2015), *Associated Press*.
[215] 200 feet.
[216] 400 miles.
[217] "Continental shelf," *National Geographic Encyclopedic Entry*.

The above suggests that the absence of human remains alongside woolly mammoths is simply due to the scarcity of the human population at the time and its concentration along now-submerged coasts. If this hypothesis is correct, it may only be a matter of time before a flash-frozen human or two is found buried next to flash-frozen mammoths.

Just because no human remains have yet been found does not, however, mean that no human being witnessed the Younger Dryas catastrophe. Indeed an account of this catastrophic event can be found far away from Siberia in the oldest known Neolithic site.

In Gobekli Tepe[218] the deepest layer dates back to ca. 10,000 BP.[219] Its major archeological feature is the 'vulture stone,' a massive carved pillar also known as *pillar 43*, shown in the picture to the right.

According to University of Edinburgh scientist Martin Sweatman,[220] the vulture stone is an astronomic representation where, like today, animals represent constellations and the whole scene displays a catastrophe caused by a cometary swarm.[221]

Computer model analysis carried out to match the patterns of the stars detailed on the Vulture Stone points to one specific date: *12,950 BP*, exactly the date of the onset of the Younger Dryas.

Figure 51: The Vulture Stone

---

[218] Major Neolithic site located in southern Turkey.
[219] Oliver Dietrich *et al.* (2013), "Establishing a Radiocarbon Sequence for Göbekli Tepe. State of Research and New Data," *Neo-Lithics* 1:36–47.
[220] Lead scientist at the University of Edinburgh and a Fellow of the Royal Society of Chemistry. Sweatman develops his extensive analysis of Gobekli Tepe in Martin Sweatman (2018), *Prehistory Decoded*, Troubador Publishing.
[221] Fiona Macdonald (2019), "Ancient Carvings Show Evidence of a Comet Swarm Hitting Earth Around 13,000 Years Ago," *Science Alert*.

Conclusion

The Younger Dryas period was marked by major catastrophes. The woolly mammoths and the Clovis people were the tragic witnesses to a major cosmic event that deeply transformed our planet only 13,000 years ago.

This event is a serious thorn in the side of uniformitarians who, despite copious evidence, still deny the facts before them. Their insistence on clinging to a uniformitarian dogma that has proven to be false can be found in the very foundation of politics and power:

> The **legitimacy of the ruling class – in whatever political form it may take – is based on the illusion that they can protect the people**, whether from war, famine, economic hardship, or any other kind of disaster that disrupts the everyday routine of their lives and livelihood.
>
> By **attributing the cause of these cosmically-induced events to humans, the elites maintain the illusion that they are, at least to some degree, in control**; if they are causing it, then theoretically, at least, they could stop it.[222]

If even a smaller version of the Younger Dryas cometary bombardment happens during our time, it will be interesting to see how the elites react, if indeed they survive to react at all. Will they acknowledge the fragile human condition and their total powerlessness in the face of cosmic forces? Or will they try to parlay the cosmically induced event into a human-made catastrophe, as they are currently doing with global warming and climate change?

While researching flash-frozen mammoths, an unexpected anomaly appeared. The Younger Dryas was a 1,400-year global cooling period which led to an overall increase in the size of ice sheets.[223] However, during this same time period,[224] sea levels *rose* by about 20 meters,[225] from 65 to 45 m[226] below the current sea level.

Cooling is associated with an increase in ice sheet volume, which leads to a *drop* in sea level, because seawater transforms into ice and atmospheric water becomes snow. Yet during the Younger Dryas, the exact opposite happened. Instead of dropping sharply, sea levels rose. This paradox leads to an obvious question: where did all this extra water come from?

---

[222] Lescaudron (2014), p. 248.
[223] Don Easterbrook (2019), "Younger Dryas," *Encyclopædia Britannica*.
[224] 12,900 BP–11,500 BP.
[225] About 57 feet.
[226] From 213 to 148 feet.

Part II:
Did Earth 'Steal' Martian Water?

Figure 52: © Myriam Kieffer
Artist depiction of an electric discharge between Mars and Earth

The Younger Dryas was a period of global cooling that lasted from 12,900 to 11,500 years ago during which surface temperatures dropped by approximately 7°C.[227]

In theory, such a severe cooling should increase the volume of polar ice – solid water – and, as a result, reduce the volume of the seas – liquid water. However, during the Younger Dryas, sea levels rose approximately 20 meters[228] over more than a millennium, as illustrated by the graph below:

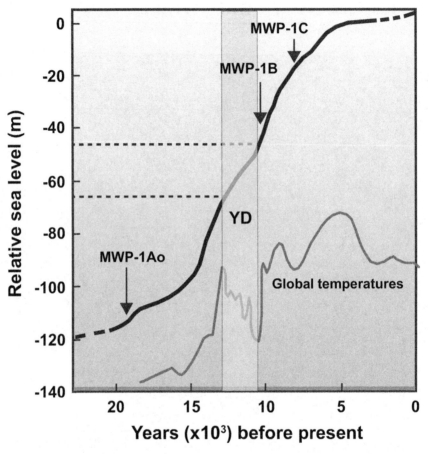

Figure 53: Sea level vs. global temperature (20,000BP–Now)

If the sea level rose while ice caps were building up, where could this water have come from?

Coincidentally or not, most of Mars' northern hemisphere was once covered with water, and this ocean has mysteriously disappeared. Where did the Martian water go? Could the mysterious appearance of water on Earth and the equally puzzling disappearance experienced on Mars be related?

---

[227] 13°F. G.S. Boulton et al. (2004), "Evidence of European ice sheet fluctuation during the last glacial cycle," *Developments in Quaternary Sciences* 2(1):441–460.
[228] 66 feet.

Sea Levels on Earth

One of the cometary impacts that triggered the Younger Dryas hit the Laurentide ice sheet.[229] This impact melted massive quantities of ice and led to a sea level rise. But, the cometary impact on the Laurentide ice sheet can only explain a small part of the 20-meter rise observed during the Younger Dryas:

> Reconstruction of the glacial melt history finds **major northward melt-water discharge 13,100–12,500 years ago**, at the beginning of the Younger Dryas. The outflow entered the Arctic Ocean, via the Mackenzie River, Fram Strait, and ultimately reached the eastern North Atlantic.
> Geomorphological data, on the other hand, suggest that ice still blocked routes to the north and east toward the St. Lawrence Seaway until the end of the Younger Dryas. Sea level curves from Tahiti, New Guinea, and Barbados **show a small step (under 6 meters) around 13,000 years ago near the Younger Dryas onset, which may have come from this deluge.**[230]

A 7°C drop in average temperatures should lead to a sea level drop of about 28 meters[231] (given a rate of -4 m/°C[232]).

To recap, we have the following three numbers:

+20 meters: observed sea level rise during Younger Dryas

+6 meters: sea level rise due to the melting of the Laurentide ice sheets

-35 meters: sea level drop that should have been induced by the Younger Dryas cooling

This means that at the time of the Younger Dryas, *about 49 meters*[233] *(20+35-6) of extra water was added to the surface of the Earth.* Keep in mind that these three figures are just estimates based on various hypotheses. They provide us nonetheless with an order of magnitude.

---

[229] See Part I, chapter "Location of Geographic North Pole before Impact."
[230] Vivien Gornitz (2013), *Rising Seas: Past, Present, Future*, Columbia University Press, p. 127.
[231] 92 feet.
[232] The change in sea level induced by global warming/cooling varies depending on the source. This figure is about 2m/°C according to the most conservative papers, see for example: Anders Levermann *et al.* (2013), "The multimillennial sea-level commitment of global warming," *Proceedings of the National Academy of Sciences* 110(34):13745–13750. According to other researchers, this figure is as high as 6 to 10m/°C. See for example: Aslak Grinsted (2013), "Relationship between sea level rise and global temperature," *Aslak Grinsted personal website.* We averaged those estimates and retained a 5m/°C number.
[233] 161 feet.

Water on Mars?

In 1666, Italian astronomer Cassini observed ice-like polar caps and clouds on Mars through simple telescope observations[234] and concluded that there was obviously water on the planet.[235]

Cassini's view prevailed for a few centuries, but modern science rejected his claim and the new doctrine became that there was no water whatsoever on Mars. It's only recently, with the massive flow of data coming from Martian probes and rovers, that the evidence became overwhelming that Mars did indeed have water at some point in the past.

According to recent estimates,[236] Mars used to hold enough water to cover its entire surface in a liquid layer about 140 meters[237] deep. About 85% of this water has, however, 'disappeared,' while the remaining 15% is stored under ice at the poles.

Apparently, Martian water was not uniformly spread over the surface of the planet. A topographic study reveals that most Martian water was stored in the north of the planet, in one single ocean, with a similar volume to Earth's Arctic Ocean.[238]

If this water was somehow transferred to Earth, it would result[239] in an approximate sea level

Figure 54: Giovanni Domenico Cassini (1625–1712)

---

[234] Telescope technology in the 17th century should not be underestimated. Cassini himself discovered four satellites of Saturn, namely Iapetus, Rhea, Tethys and Dione, through his telescope observations. See: Albert Van Helden (2009), "The beginnings, from Lipperhey to Huygens and Cassini," *Experimental Astronomy* 25 (1–3).

[235] David Harland (2007), *Water and the Search for Life on Mars*, Springer Science & Business Media, p. 3.

[236] G. L. Villanueva et al. (2015), "Strong water isotopic anomalies in the Martian atmosphere: Probing current and ancient reservoirs," *Science* 348(6231).

[237] 460 feet.

[238] Bill Condie (2015), "Mars lost more water than the volume of the Arctic Ocean," *Cosmos Magazine*.

[239] The volume of water that Mars once held is estimated to have been about 20,000,000 km^3 (about 5,000,000 cubic miles). 85% of this volume is 17,000,000 km^3 (about 5,000,000 cubic miles). A 1-meter increase in sea level requires about 400,000 km^3 according to Bethan Davies (2020), "Calculating glacier ice volumes and sea level equivalents," *Antarctic Glaciers*. Therefore, 85% of Mars' water should have induced a 42 meters (17,000,000 / 400,000) sea level rise on Earth.

rise of *42 meters*.[240] Coincidently or not, this figure is comparable, in terms of magnitude, to the hypothesized 49 meters[241] of 'external' water received by Earth during the Younger Dryas.

Figure 55: © Laser Atelier
Topographic map of Mars with its ocean

---

[240] 140 feet.
[241] 161 feet.

How Could Mars Lose Its Water?

As noted previously, most water on Mars has 'disappeared.' Modern science offers two explanations for this: underground leakage and space leakage.

Underground leakage is highly unlikely because Mars has no known tectonic plates[242] and therefore no subduction, which is the main process by which surface water is brought underground,[243] as shown by the drawing below, where the oceanic water represented in black sinks in a subduction channel below the crust:

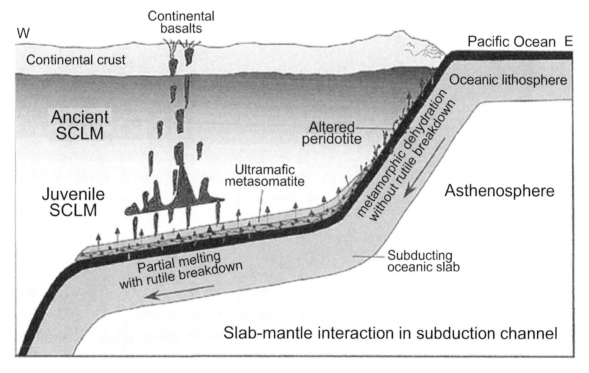

Figure 56: © Zheng Xu
Subduction of oceanic water below continental crust

Space leakage posits that, about four billion years ago, Mars lost its magnetic field[244] and, devoid of this protection, solar winds stripped the planet of its atmosphere, and most of its water, within a few hundred million years.[245]

---

[242] Michael Carr (2006), *The Surface of Mars*, Cambridge University Press, p. 16.
[243] C. Cai *et al.* (2018), "Water input into the Mariana subduction zone estimated from ocean-bottom seismic data," *Nature* 563:389–392.
[244] Anna Mittelholz *et al.* (2020), "Timing of the martian dynamo: New constraints for a core field at 4.5 and 3.7 Ga," *Science Advances*.
[245] Mike Wall (2015), "Mars Lost Atmosphere to Space as Life Took Hold on Earth," *Space.com*.

If, as posited by the space leakage theory, Mars' ocean disappeared about four billion year ago, i.e. soon after Mars' formation estimated to have occurred 4.5 billion years ago,[246] there should be roughly the same density of craters on the oceanic bed as outside of it.

However, this theory doesn't hold water (pun intended) for one simple reason: the top half of the Martian northern hemisphere (where the Martian ocean once stood) exhibits *far fewer and far smaller craters than the rest of the planet*.

The picture[247] below reveals that Mars' oceanic bed (Vastitas Borealis) is *almost devoid of craters*, while the Martian continent (Arabia Terra and Terra Sabaea) is littered with them:

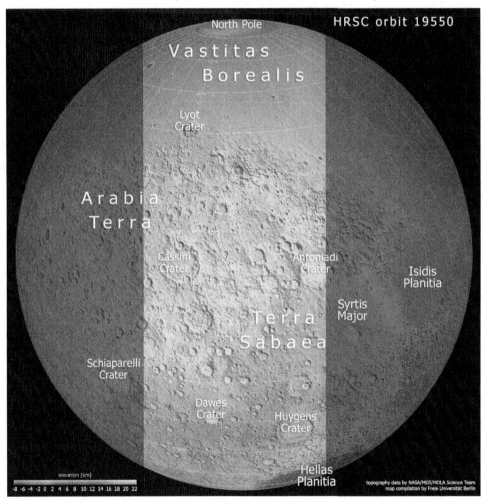

Figure 57: © NASA
Mars topographic map highlighting cratered and pockmarked swathes

---

[246] Lucy-Ann McFadden *et al.* (2006), *Encyclopedia of the Solar System*, Elsevier, p. 318.
[247] This colour-coded topographic image shows a slice of the Red Planet from the northern polar cap downwards, and highlights cratered, pockmarked swathes of the Terra Sabaea and Arabia Terra regions. The area outlined in the centre of the image indicates the area imaged by the Mars Express High Resolution Stereo Camera on 17 June 2019 during orbit 19550. See: European Space Agency (2019), "From clouds to craters: Mars Express," *Phys.org*.

If Mars' ocean disappeared about four billion years ago, how can we explain that the Martian ocean bed is almost devoid of asteroid impact evidence while the rest of the planet is covered with craters?

One potential explanation would be that most impacts on Mars happened more than 4 billion years ago, when the ocean was still there and acted as a damper, preventing the formation of craters on the oceanic bed.

However, this explanation doesn't seem to hold up. Despite an almost non-existent atmosphere, violent dust storms do occur on Mars,[248] as illustrated by the following picture:

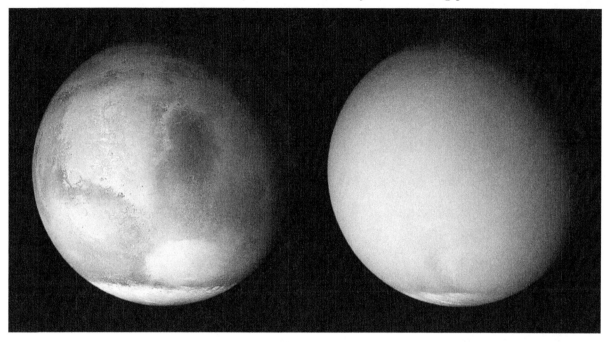

Figure 58: © NASA/JPL/MSSS
Left: Mars without a dust storm (06/2001) Right: Mars with a global dust storm (07/2001)

A consequence of those global dust storms is that Martian craters are subjected to intense erosion. Therefore, the well-preserved craters must be relatively recent.

In 2011, Robbins *et al.*[249] finalized a database listing close to 400,000 Martian craters. The map below is extracted from this database and shows the geographic distribution of well-preserved, recent craters where the lighter the square, the higher the proportion of recent craters. As you can see, there are more light grey squares in the northern hemisphere, i.e. there is a higher proportion of recent craters where the Martian ocean used to be.

The almost pristine Martian oceanic bed combined with the high proportion of recent craters where Mars' ocean once stood strongly suggests that Mars lost its ocean *much more recently than usually claimed by mainstream science.*

---

[248] C. E. Leovy *et al.* (1973), "Mechanisms for Mars Dust Storms," *Journal of the Atmospheric Sciences* 30(5):749–762.

[249] S. J. Robbins *et al.* (2012), "A new global database of Mars impact craters ≥1 km: 2. Global crater properties and regional variations of the simple-to-complex transition diameter," Journal of *Geophysical Research* 117(E6).

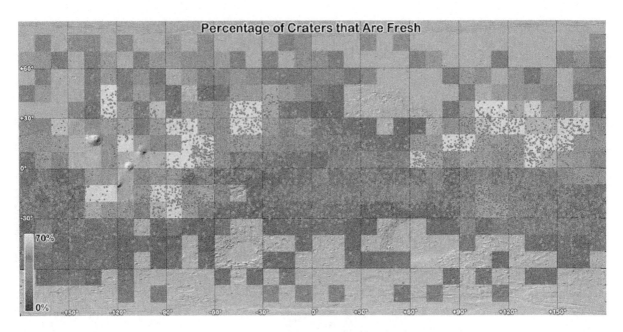

Figure 59: © Robbins
Geographic concentration of well-preserved craters on Mars

Interplanetary Electric Discharge

The Electric Universe theory, as described in my previous book,[250] shows how celestial bodies like planets, stars, moons or comets are electrically charged. In addition, such bodies are surrounded by a sort of 'insulation bubble,'[251]

As a general rule astronomical bodies hold different electrical charges, therefore when two astronomical bodies – two planets for example – get close enough, an electron discharge forms from the most negative planet to the most positive one, in order to balance the electric charge difference. This phenomenon has been observed several times. Here are a few examples:

Comet Shoemaker-Levy and Jupiter

Back in July 1994, Comet Shoemaker-Levy fragmented on its way to Jupiter. This is what happened to one of those fragments:

Figure 60: © Siding Spring Observatory July 16, 1994. Electric discharge between Jupiter and comet Shoemaker-Levy

> The Hubble Space Telescope detected a flare-up of fragment "G" of Shoemaker-Levy long before impact at a distance of **2.3 million miles**[252] **from Jupiter.** For the electrical theorists this flash would occur as the fragment crossed Jupiter's plasma sheath, or magnetosphere boundary.[253]

The pre-impact electric discharge was so intense that scientists measured record levels of radio emissions coming from Jupiter. As testified by Michael Klein of JPL[254]:

> Never in 23 years of Jupiter observations have we seen such a rapid and intense increase in radio emission.[255]

The picture to the right confirms the magnitude of the electric discharge.

---

[250] Lescaudron (2014), Chapters 7 and 8.
[251] Also known as "double layer" or "Langmuir sheath." For more details see: Stephen Smith (2020), "Langmuir Sheaths," *Thunderbolts.info*.
[252] 3.7 million km.
[253] David Talbott *et al.* (2006), "Deep Impact and Shoemaker-Levy 9," *Thunderbolts.info*.
[254] Jet Propulsion Laboratory.
[255] Wal Thornhill (2005), "Comet Tempel 1's Electrifying Impact," *Holoscience.com*.

Jupiter and one of its moons: Io

Io exhibits a puzzling bluish plume that rises hundreds of kilometers above its surface.[256] This plume has been constantly active since its first observation by the Voyager space probe in 1979.[257] The official explanation for this phenomenon is volcanism, which is unlikely considering the duration, height and shape of the plume. According to several scientists, Io's plume is evidence of intense electrical discharging:

> In November 1979, the noted astrophysicist Thomas Gold[258] proposed that the gigantic plumes on Io are not volcanic but **evidence of electrical discharging**. Years later, a paper[259] by Peratt and Alex Dessler followed up Gold's suggestion, showing that the **discharges took the form of a 'plasma gun effect,'** which produces a parabolic plume profile, filamentation of the matter within the plume, and the termination of the plume onto a thin annular ring.[260]

Figure 61: © NASA/JPL
Io exhibiting a massive electrical discharge

---

[256] David Talbott *et al.* (2004), "Electric Jets on Io," *Thunderbolts.info*.
[257] NASA Content Administrator (2008), "Io: The Prometheus Plume," *Nasa.gov*.
[258] Thomas Gold. (1979), "Electrical Origin of the Outbursts on Io," *Science* 206(4422):1071–1073.
[259] A. Peratt *et al.* (1988), "Filamentation of Volcanic Plumes on Io," *Astrophysics and Space Science* 144:451–456.
[260] W. Thornhill *et al.* (2007), *The Electric Universe*, Mikamar Publishing, p. 112.

Herbig Haro Objects

Herbig Haro Objects are bright, high, ionized regions of space containing 'young' stars and planets.[261] Electric discharges in the form of gigantic interstellar Birkeland currents occur between proto stars and proto planets, and provide Herbig Haro objects with their brightness[262]:

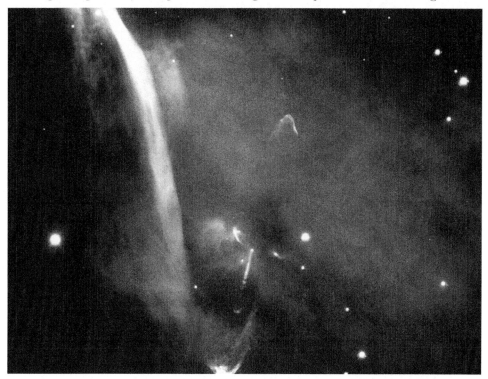

Figure 62: © ESO
Electric discharge along HH34 celestial bodies

Earth and the Moon

Electrical discharges between astronomical bodies even occur between planet Earth and its Moon. Those discharges are less intense than the previous example but they are nonetheless very real:

> The Lunar Prospector spaceship detected changes in the lunar nightside voltage during [Earth's] magnetotail crossings, **jumping from −200 V to −1000 V**[263].

Once every lunar month, the Moon crosses the Earth's magnetotail and an electric deluge occurs on our satellite, which is literally bombarded with a variety of ions.

---

[261] K. P. M. A. Blagrave et al. (2006), "Photoionized Herbig-Haro Object in the Orion Nebula," *The Astrophysical Journal*.
[262] Theodore Holden et al. (2019), *Cosmos In Collision: The Prehistory of Our Solar System and of Modern Man*, Lulu Press, Inc.
[263] Jasper Halekas et al. (2008), "Lunar Prospector observations of the electrostatic potential of the lunar surface and its response to incident currents," *Journal of Geophysical Research: Space Physics* 113(A9).

Data from Kaguya[264] now suggests that some of those ions are oxygen and about 26,000 of them hit every square centimeter[265] of the Moon's surface every second during this electric deluge.[266]

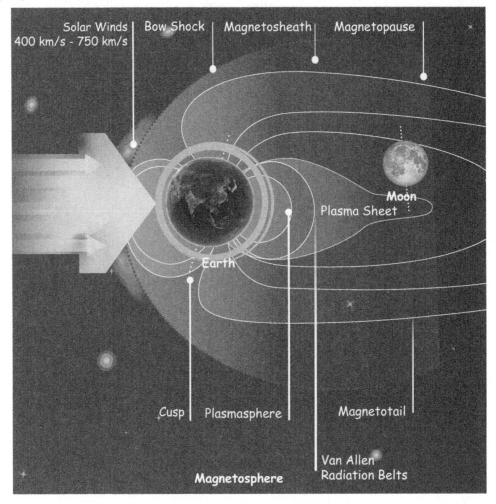

Figure 63: © Design context
The Moon crossing the Earth's magnetotail

Electric discharges between celestial bodies are similar to arc welding, the main difference being the scale.

When the negatively charged electrode is brought close enough to the positively charged part, an electric arc forms and electrons travel through the ionized air (plasma) along what is called a "Birkeland current." Electrons travel from the electrode (stick) to the metal in order to re-balance the electric charges.

Notice that during arc welding, *electrons are not the only material transferred from the electrode to the metal*; (negatively charged) molten metal from the tip of the electrode is carried along with

---

[264] Japan's second spacecraft to explore the Moon from orbit.
[265] 0.155 square inch.
[266] Kentaro Terada *et al.* (2017), "Biogenic oxygen from Earth transported to the Moon by a wind of magnetospheric ions," *Nature Astronomy*.

electrons towards the positively charged metal.

A typical feature of electric discharges is 'electric scarring.' These fractal patterns are also known as Lichtenberg figures.[267] Notice that the polarity of the scarred material has a notable influence on the form of the Lichtenberg figure:

> ... there is also a marked difference in the form of the figure, according to the polarity of the electrical charge that was applied to the plate. If the charge areas were **positive**, a widely extending patch is seen on the plate, consisting of a **dense nucleus, from which branches radiate in all directions.**
>
> **Negatively charged areas are considerably smaller and have a sharp circular or fan-like boundary entirely devoid of branches.**[268]

Lichtenberg used a mixture of yellow sulfur powder, which is attracted to positive charges, and red lead powder, which is attracted to negative charges. He subjected these powders to high-voltage electric discharges.[269] The two following photographs show the result of this experiment:

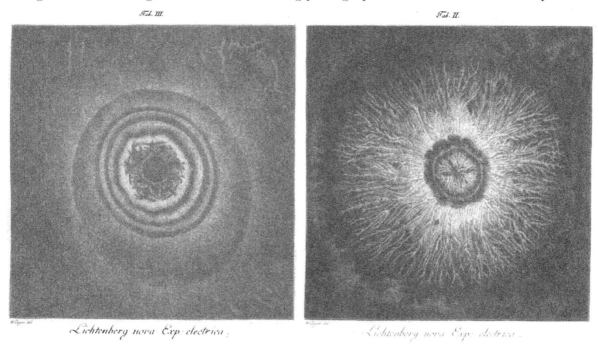

Figure 64: © Lichtenberg
Left: Positive Lichtenberg figure. Right: Negative Lichtenberg figure

Let's keep in mind the shape of the Lichtenberg figures because we will soon encounter them again,[270] albeit on a much larger planetary scale.

---

[267] Georg Christoph Lichtenberg (1742–1799) was a German physicist, satirist, and anglophile. Lichtenberg discovered the "Lichtenberg figures" in 1777.
[268] Heinrich Hertz (1900), *Electric Waves: Being Researches on the Propagation of Electric Action with Finite Velocity Through Space*, Macmillan and Company.
[269] Stoneridge Engineering, "What are Lichtenberg figures, and how do we make them?" (last update 2019), *Capturedlightning.com*.
[270] See chapter "Signs of electric discharge on Mars" in this second part.

Relative Polarity of Mars and Earth

As described in my previous book,[271] the Sun is the most positive body within the solar system. Therefore, the further away from the Sun a planet is, the more negative its electric potential is. Being further away from the Sun than Earth, Mars' electric potential is lower than that of Earth.

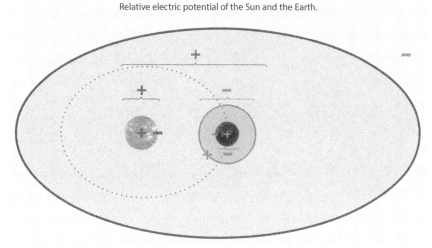

Figure 65: © Sott.net
Sun, Earth, heliopshere: relative electric charges

As a consequence, *if* an electric discharge happened between Mars and Earth, it started from the most negatively charged body (Mars) and spread towards the most positively charged body (Earth). Mars was the cathode (negatively charged) and was stripped of material (gases, rocks, water), and its electric scarring should exhibit craters, striking at a high point, forming additional craters and steep-sided trenches.

> If the surface is a cathode (negatively charged), the arc will tend to move across the surface. After striking, usually at a high point, and eroding a crater, the arc may jump to a new high point – the rim of the new crater is a most likely target.
>
> The abundance [on Mars] of small craters centered on the rims of larger ones testifies to this predictable behavior. As the arc travels, it may erode a series of craters in a line, appearing as a chain of craters.
>
> If the craters in these chains overlap, the effect is a steep-sided trench with scalloped edges. The arc may erode a trench for a distance and then jump some distance away before eroding another trench. These 'dashed line' trenches will usually have circular ends and constant widths. **All of these patterns occur in great abundance on the surface of Mars.**[272]

---

[271] Lescaudron (2014), chapter 8.
[272] Wal Thornhill (2005), "The Electric Universe: Part II. Discharges and Scars," *Thunderbolts.info*.

Signs of Electric Discharge on Mars

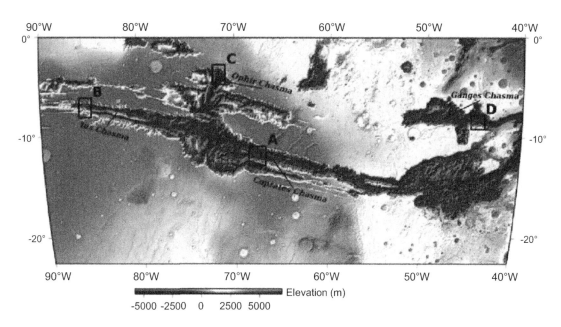

Figure 66: © MOLA database
Topographic map of Valles Marineris

If a massive electric discharge occurred between Mars and Earth, is there any trace of a major (negative) Lichtenberg figure, as described above, to be found on Mars?

One of the main geological features of Mars is Valles Marineris. At more than 3,000 km[273] long, 600 km[274] wide and up to 8 km[275] deep, it is the largest canyon in the entire solar system, and *stretches for nearly a quarter of the planet's circumference.*[276]

Several researchers theorized that Valles Marineris formed as a result of water erosion billions of years ago.[277] However, this explanation doesn't seem to match some of its characteristics:

---

[273] 1,800 miles.
[274] 360 miles.
[275] 5 miles.
[276] NASA Content Administrator (2008), "Valles Marineris: The Grand Canyon of Mars," *Nasa.org*.
[277] See for example: Devon Burr *et al.* (2009), *Megaflooding on Earth and Mars*, Cambridge University Press, p. 214.

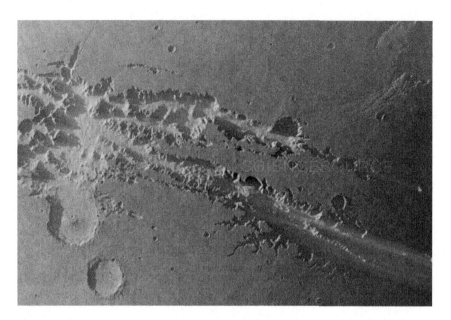

Figure 67: © Science Photo Library
Probe photograph of Valles Marineris

- In Valles Marineris, the 'outflow' is as narrow as the 'inflow,' and the middle of the course constitutes its broadest part. Overall the width is quite constant, unlike rivers, which tend to broaden over their course.

- The 'course' of Valles Marineris doesn't follow the down slope. It sometimes 'runs' uphill although there is no sign of the damage – rifts for example – that might be expected if the topographical changes were due to later vertical movement of the terrain.

- Valles Marineris doesn't reveal signs of tributary 'rivers.' The two major 'rivers' that can be imagined run parallel to each other. The secondary 'river' joins the main one at a near right angle, unlike the converging path usually exhibited by tributaries that join a main river.

- The bed of Valles Marineris reveals transverse markings, unlike riverbeds, which tend to have longitudinal markings shaped by the river flow.

- The 'tributaries' exhibit a V-shaped cross-section while water erosion typically forms U-shaped riverbeds.

- Valles Marineris' banks are very deep, up to 7 km,[278] and very steep. The banks show no sign of water erosion and its typical horizontal marking. On the contrary, the markings reveal a vertical chevron pattern.

---

[278] 4.4 miles.

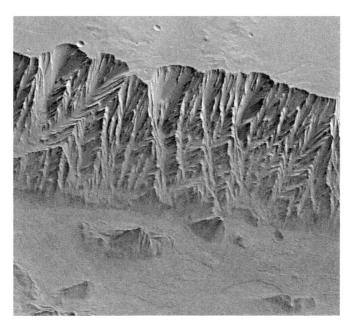

Figure 68: Bank of Valles Marineris

While the features of Valles Marineris seem to contradict the water erosion theory, they are very consistent with the distinctive features of (negative) electric scarring:

> When planets come close, gargantuan interplanetary lightning results. It is **perfectly capable of stripping rock and gases from a planet against the puny force of gravity. It does so leaving characteristic scars.** ...
> The parallelism of the canyons is due to the long-range magnetic attraction of current filaments and their short-range strong electrostatic repulsion.
> Particularly significant are the small parallel rilles composed essentially of chains of craters. A traveling underground explosion follows the lightning streamer and cleanly forms the V-shaped tributary canyons.
> There is no collapse debris associated with undercutting water flow. Similarly, the "V" cross-section is usual for craters formed by underground nuclear explosions. The circular ends of the tributaries, where the explosion began, are precisely of that shape.
> In comparison, headward erosion by ground water sapping gives a U-shaped cross-section and does not necessarily end in a circular alcove.
> Note that some of the tributary canyons on the south rim of Valles Marineris cut across one another at near right angles. This might be due to repeated discharges from the same area chasing the main stroke as it travelled along Ius Chasma. No form of water erosion can produce crosscutting channels like that.
> **The fluted appearance of the main canyon walls is probably due to the same travelling explosive action.**[279]

Interestingly, *Valles Marineris is contiguous to the ocean that once covered most of Mars' ocean.* If Valles Marineris was the location of an electric discharge between Mars and Earth, the adjacent Martian ocean would have certainly been affected, and possibly transferred.

---

[279]Wal Thornhill (2001), "Mars and the Grand Canyon," *Holoscience.com*.

Evidence of Material Transfer from Mars to Earth

As mentioned in the above quote, a massive electric discharge from Mars to Earth could have stripped significant quantities of rocks from Valles Marineris. So, before looking for signs of a major (positive) electric discharge on Earth, let's see if there is any evidence of Martian rocks on Earth. As of 2019, *237 Martian meteorites have been found on Earth*, according to the Meteoritical Society.[280] So, transfer of material from Mars has occurred.

One might assume that this phenomenon is very ancient and happened billions of years ago when the planets were forming, and when asteroids were rampant and orbits were unstable. But data suggests that this is not exactly the case.

While the landing time for most Martian meteorites is unknown, a few have been dated – in particular the Martian meteorite commonly abbreviated ALH84001.[281] Its ETA on Earth has been estimated at *13,000 years ago*.[282].

ALH84001 is an orthopyroxenite[283] meteorite and the only place on Mars where high-resolution spectral analysis has revealed the presence of orthopyroxenite is Valles Marineris.[284] Consequently, *ALH84001 is likely to come from Valles Marineris.*

In fact, ALH84001 is the only orthopyroxenite Martian meteorite. No other meteorite of this kind has ever been found on Earth.

Interestingly, because of its specific carbonate content, *ALH84001 is the only meteorite originating from a time period during which Mars is suspected to have supported liquid water.*[285]

Figure 69: © NASA
Martian meteorite ALH84001

ALH84001 is an abbreviation which stands for Allan Hills 84001. Allan Hills is the place where ALH84001 was discovered, located along the southern coast of Antarctica.

Now, let's recapitulate some key characteristics of ALH84001:

- It comes from Valles Marineris.

---

[280]Meteoritical Society – International Society for Meteorites and Planetary Science, "Meteoritical Bulletin Database," *Lpi.usra.edu*.
[281]ALH84001 is a fragment of Martian meteorite which weighs about 2 kg (4.4 pounds) and was discovered in 1984.
[282]Lunar and Planetary Institute Editors (2014), "How could ALH84001 get from Mars to Earth?" *Lunar and Planetary Institute*.
[283]Pyroxenes are group of inosilicate minerals. Their general formula is $XY(Si,Al)_2O_6$, where X represents elements like calcium, sodium or magnesium while Y represents smaller elements like chromium, aluminium or magnesium. See: Hobart King (2020), "The Pyroxene Mineral Group," *Geology.com*.
[284]David Chandler (2005), "Birthplace of famous Mars meteorite pinpointed," *New Scientist*.
[285]Matt Crenson (2006), "After 10 years, few believe life on Mars," *USAToday*.

- At the time of its arrival on Earth, Mars was a wet planet.

- It landed on Earth 13,000 years ago.

- It was found in Antarctica.

It would be interesting to know if some Martian meteorites come from Mars' oceanic bedrock. Unfortunately, its geological composition is unknown because it is covered with a thick layer of sediment.[286] However the mineralogical composition of the coast of Mars' dry ocean is known and *directly related to some Martian meteorites found on Earth.*

Indeed, there is a rare type of Martian meteorite called 'nakhlite.' Only 21 specimens have been found on Earth so far.[287] Nakhlites are rich in augite, a silicium-based mineral, and they formed from basaltic magma about 1.3 billion years ago.[288]

Because of the composition and age of the nakhlites, they are believed to originate from one of these three Martian volcanic areas: Tharsis, Elysium, or Syrtis Major Planum.[289] Coincidently or not, *each of those three volcanic constructs is situated near the coast* of what was once the Martian ocean, as shown in the following map:

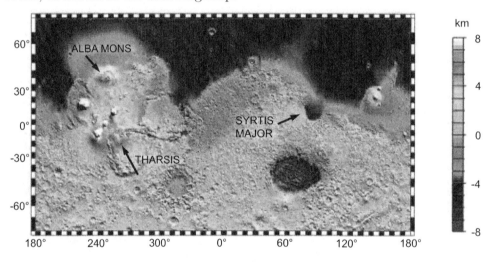

Figure 70: © NASA / GSC
Theorized geographical origin of nakhlites

Now, of the 21 nakhlite meteorites known to have reached Earth, 7 were found in Antarctica – that is 33%. This is a high percentage considering that only about 12% of all known meteorites that reached Earth have been found in Antarctica. Mass wise, 16.9 kg[290] of nakhlite meteorites were found in Antarctica – that is 54% of the total mass of nakhlite meteorites. The above suggests a tropism of nakhlite Martian meteorites towards Antarctica.

---

[286] M.R. Salvatore, *et al.* (2015), "On the origin of the Vastitas Borealis Formation in Chryse and Acidalia Planitiae, Mars," *JGR Planets* 119:2437–2456.

[287] "Meteoritical Bulletin Database," query: "Martian (nakhlite)."

[288] A.H. Treiman (2005), "The nakhlite meteorites: Augite-rich igneous rocks from Mars," *Geochemistry* 65(3):203–270.

[289] Ibid., p. 259.

[290] 37 pounds.

Lastly, the nakhlite meteorites are believed to have fallen to Earth about 10,000 years ago.[291] This figure is quite close to the arrival date of ALH84001 (13,000 years ago).

Any Sign of Electric Discharge on Earth?

If a massive electric discharge initiated from Valles Marineris and hit Earth, where did the hit possibly occur? There are several canyons on Earth, including the Grand Canyon, which hold features of electric scarring.[292] However, the data about Martian meteors provided above reveals a strong affinity of Martian meteorites for Antarctica.

So, does the bedrock of Antarctica show any sign of positive electric scarring like a massive canyon-like geological feature? Until recently the answer was 'no.' But a satellite spectroradiometry survey conducted in 2016 allowed us to see through kilometers of ice sheet and revealed that Antarctica could actually host the largest canyon on Earth:

> ... the largest unsurveyed region on the icy continent is a region called Princess Elizabeth Land. Now a team of geologists has scoured that area to reveal a massive subglacial lake and a **series of canyons, one of which – more than twice as long as the Grand Canyon – could rank as Earth's largest.**[293]

The image on the next page (Figure 71) shows the size, location and shape of the newly discovered Antarctic canyon.[294]

At this point, Martian meteorites and traces of electric scarring point to Antarctica for a potential location for the Mars–Earth transfer. But what about the main constituent of the whole process: water?

If Mars lost most of its water to Earth, there should be some evidence of this massive transfer, on our planet in general and in Antarctica in particular.

Well, Antarctica is covered by a massive volume of frozen water. Could part of the Antarctic ice sheet be of Martian origin? To answer this question, let's first observe the Antarctic ice sheet (Figure 72) and then we will compare it to its Arctic counterpart.

The Antarctic ice sheet is massive, its total volume about 30 million km^3 of ice.[295] This represents more than 70% of Earth's freshwater.[296] In comparison, the Arctic ice sheet, located over Greenland, is only 2.9 million km^3.[297]

So, in terms of volume, the northern ice sheet is less than 10% of Antarctica's ice sheet. Notice also that Antarctica does not form a single, solid continent. It's more like an archipelago comprising a few massive islands separated by deep marine areas, as depicted in the map below.

---

[291] Ibid., p. 204.
[292] Michael Armstrong (2008), "The Grand Canyon: Part One," *Thunderbolts.info*.
[293] Shannon Hall (2016), "The World's Grandest Canyon May Be Hidden beneath Antarctica," *Scientific American*.
[294] In an earlier section we described the pattern of a positive discharge based on Lichtenberg's work. Unfortunately, the existing radar images don't allow for a close examination to determine if the scarring in the Antarctic conforms to Lichtenberg's figures.
[295] 7.2 million cubic miles.
[296] American Museum of Natural History (2016), "Where is Earth's Water?" *Amonh.org*.
[297] 0.7 million cubic miles. Alexandra Witze (2008), "Climate change: Losing Greenland," *Nature* 452:798–802.

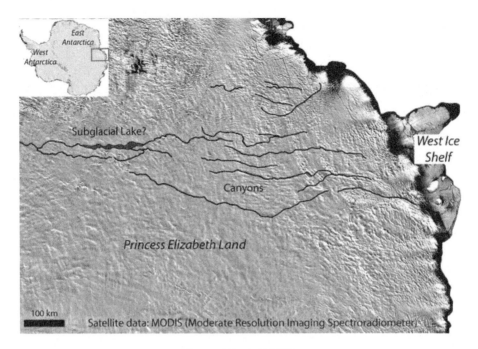

Figure 71: © MODIS
Spectroradiometry of Princess Elizabeth Land

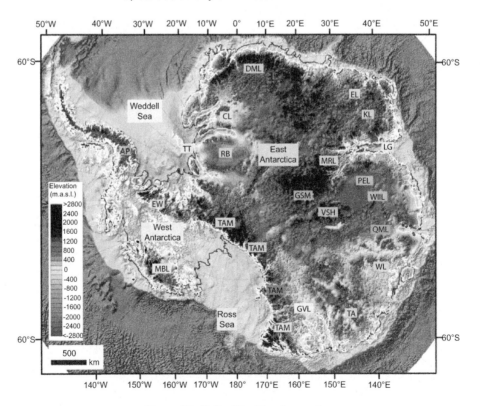

Figure 72: © BedMachine Antarctica
Topographic map of Antarctica

As shown by the Antarctica cross section below, in some places, like the one indicated by the vertical dashed line, the Antarctic ice sheet is almost 4 km[298] thick, *extending from the bedrock more than 1.5 kilometers*[299] *below sea level* up to the top of the ice sheet, which is more than 1.5 km above sea level:

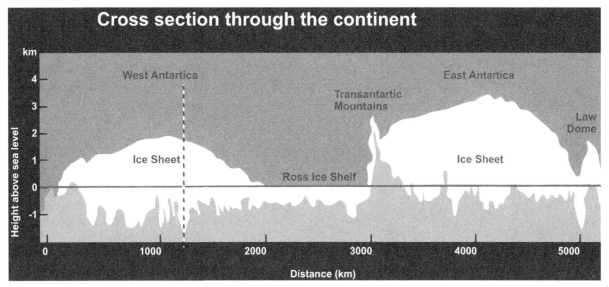

Figure 73: © Discovering Antarctica
Antarctic cross section

For comparison, the Arctic sea ice is about one thousand times thinner,[300] reaching a maximum thickness of 4 meters[301] with ridges up to 20 meters,[302] although the average depth of the Arctic Ocean is about one thousand meters,[303] which is comparable to the average depth of the Antarctic bedrock.

In light of these huge differences, an obvious question arises: *why is there so much more ice in Antarctica compared to the Arctic?* Why does Antarctic ice extend more than 1,500 meters below sea level and reach the bedrock, while Arctic ice is a mere 4-meter-thick layer floating on the ocean?

According to mainstream science the Antarctic and Greenland ice sheets were formed by the same process: the incremental accumulation of snow, year after year. This suggests that Antarctica experienced a lot more snowfall. But data shows the opposite. Indeed, Antarctica is one of the driest places on Earth, with precipitation of only 16 cm/year,[304] while the Arctic basin experiences higher precipitation with more than 20 cm/year[305] and Greenland reaches a whopping

---

[298] 2.7 miles.
[299] 1 mile.
[300] Michon Scott *et al.* (2016), "Sea Ice," *NASA Earth Observatory.*
[301] 13 feet.
[302] 66 feet.
[303] Ned Ostenso (1998), "Arctic Ocean," *Encyclopædia Britannica.*
[304] About 6 inches. David Vaughan *et al.* (1999), "Reassessment of Net Surface Mass Balance in Antarctica," *Journal of Climate* 12(4):933–946.
[305] About 8 inches. M.C. Serreze *et al.* (2000), "Representation of Mean Arctic Precipitation from NCEP–NCAR

40 to 120 cm/year.[306]

If Antarctica receives less snow than the Arctic region, the only explanation for its tenfold higher quantity of ice is that it experiences less melting. So, Antarctica must have repeatedly experienced much colder temperatures relative to the Arctic region. Again data suggests the opposite.

For eons the Arctic region has been much colder than Antarctica. It is only during the last 11,000 years that the Antarctic has been marginally colder than the Arctic, as shown in the graph below:

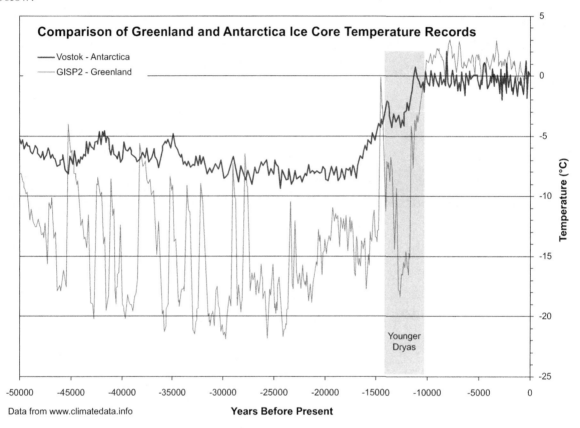

Figure 74: © Climatedata.info
Arctic vs. Antarctic temperatures

The diagram above shows that around 11,000 BP, a *sudden and marked de-correlation* between Greenland (GISP2) and Vostok (Antarctica) ice core–based temperature reconstruction occurred.

Between 11,000 BP and now, the two temperature curves are very similar in shape and very close in terms of value. This is to be expected for two ice caps both located on the same planet and, therefore, exposed to virtually the same temperature variations.

But before this time (50,000 BP to 11,000 BP) the two curves are totally different – so different

---

and ERA Reanalyses," *Journal of Climate* 13(1):182–201.

[306] 16 to 47 inches a year. M.C. Serreze, and R.G. Barry (2005), *The Arctic Climate System*, Cambridge University Press.

and de-correlated that one might wonder if these two curves are telling us the story of the environmental conditions of *two different planets*. Is the Antarctica ice older than 11,000 BP actually Martian ice featuring therefore the climatic variations of planet Mars?

In any case, no endogenous cause – difference in snowfall or temperature – can explain the marked difference in depth and volume between the Antarctic ice sheet and the Arctic one. *A massive and sudden inflow of exogenous water, in ice form,*[307] *dumped in Antarctica would, however, explain these discrepancies.*

Notice that the estimated volume of Antarctica is 27 million km^3 of ice.[308] Actually this number is consistent with the volume of Mars' lost ocean: 17 million[309] km^3, complemented with precipitation over Antarctica during the past 11 millennia or so.

In conclusion, the factors exposed above relative to the features of Martian meteorites, the electric scarring in Valles Marineris and Antarctica's bedrock, the peculiarities of the Antarctic ice sheet and Mars' lost ocean *all point out to an electrically driven transfer of material, including rock and ice, from Mars to Earth.*

Actually the transfer of massive quantities of water to Earth during meteoritic events is not totally foreign to mainstream science:

> Our calculations indicate that, coinciding with the so-called Heavy Bombardment produced by the gravitational destabilization of the main asteroid belt, **billions of tons of carbonaceous chondrites reached Earth** about 3,800 million years ago. **And they did it transporting in their fine matrices water** and other volatile elements in form of hydrated minerals.[310]

Of course, mainstream science only acknowledges catastrophes if they have occurred millions or (even better) billions of years ago, but nonetheless massive transfer of rock and water from space to Earth is a reality.

---

[307] In space water boils then freezes, therefore the form of water in space is solid (ice). For more information about this topic, see: Ethan Siegel (2014), "Does water freeze or boil in space?" *Medium.com*.
[308] 950 cubic feet. Davies (2020).
[309] 600 cubic feet.
[310] Josep M. Trigo-Rodríguez *et al.* (2019), "Accretion of Water in Carbonaceous Chondrites: Current Evidence and Implications for the Delivery of Water to Early Earth," *Space Science Reviews*.

How Could Mars Get So Close to Earth?

Mars exhibits the second largest eccentricity of all planets in the solar system. Large eccentricities usually suggest orbits that were disrupted in the recent past.[311] Because of this marked eccentricity, Mars can get as close as 56 million km[312] from Earth,[313] as shown in the diagram below:

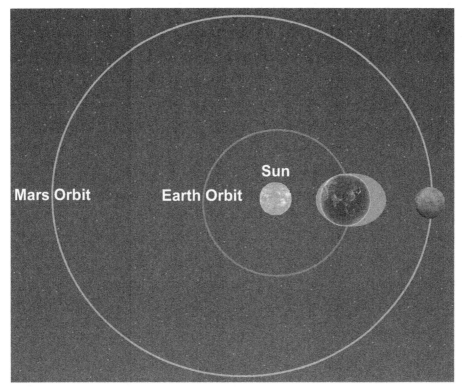

Figure 75: © Sott.net
Mars and Earth orbits

For comparison, Earth's magnetotail[314] extends more than 6 million km,[315] as symbolized by the white ellipse around Earth in the image above.

So, electrically speaking, Mars is only one order of magnitude away from of Earth. However, the current distance between Earth and Mars is too large for any electric discharge to occur

---

[311] Ian Garrick-Bethell *et al.* (2006), "Evidence for a Past High-Eccentricity Lunar Orbit," *Science* 313(5787):652–655.
[312] 37 million miles.
[313] Tony Phillips, (2003), "Approaching Mars," *Nasa.gov*.
[314] Jake Port (2016), "What causes an aurora over the poles?" *Cosmos Magazine*.
[315] 3.7 miles.

between the two planets. But could some kind of cosmic disruption have brought the two planets abnormally close?

The obvious agent for such a massive orbital disruption would be a comet, *but a very large one* – large enough to move Mars away from its initial orbit.[316]

This scenario is actually the main theory developed by Immanuel Velikovsky[317] in his best-selling book *Worlds in Collision*.[318]

Figure 76: © Frederic Jueneman Immanuel Velikovsky in 1974

Using mostly comparative mythology, Velikovsky proposed that *Venus was initially a comet that disrupted the orbit of Mars, which then made a close approach to Earth*, hence the material transfer described above.

Scientific leaders ruthlessly lambasted Velikovsky's catastrophist theory because it directly threatened their fundamental paradigm, *uniformitarianism*, without which the church of materialist progress and its Darwinian atheist creed would inevitably collapse. Adding insult to injury, Velikovsky based his work on religious texts and showed that they might carry more scientific data than previously believed.

Velikovsky realized that if his scenario was true, *several predictions could be made about the astronomical bodies involved*. After all, the merit of a theory is based on its predictive abilities. The predictions Velikovsky made were in total contradiction with the prevailing views of the time.

Decade after decade, space programs provided data that made it possible to test Velikovsky's claims. Unexpectedly, most of them turned out to be true. Some of the most noticeable predictions were the Jupiter radio signal,[319] the Sun net electric charge[320] and Earth's magnetosphere extending beyond the Moon.[321]

Analyzing all the accurate predictions made by Velikovsky is beyond the scope of this book. However, having already gathered information about a potential encounter between Mars and Earth, we will now focus on the last piece of the puzzle: Venus, and, in particular, Velikovsky's predictions relative to its cometary nature.

The nature of Venus was the pivotal point of the controversy surrounding *Worlds in Collision*. If Venus was not a comet, the whole chain of events was impossible. Conversely, if Venus was indeed a comet, Velikovsky's Earth–Mars close encounter scenario becomes much more plausible.

---

[316] For comparison Mars is about ten times heavier than the Moon and ten times lighter than Earth. See: Tim Sharp (2012), "How Big is Mars? Size of Planet Mars," *Space.com*.

[317] A Princeton scholar, Velikovsky (1895–1979) wrote several books developing groundbreaking interpretations of ancient history and myth. Before that, Velikovsky played an important role in the founding of the Hebrew University of Jerusalem. He was also a psychoanalyst, psychiatrist and psychoanalyst.

[318] Immanuel Velikovsky (1950), *Worlds in Collision*, Macmillan Publishers. The book was an instant *New York Times* bestseller, topping the charts for eleven weeks and remaining in the top ten for twenty-seven weeks straight.

[319] V. Bargmann *et al.* (1962), "On the Recent Discoveries Concerning Jupiter and Venus," *Science* 138(3547):1350–1352.

[320] "Electric Sun Model," *The Velikovsky Encyclopedia*.

[321] "Worlds in Collision," *The Velikovsky Encyclopedia*.

Was Venus a Comet?

According to mainstream science, Venus is a sister planet of Earth and Mars. They formed the same way, via accretion, from the same material, in the same region, over the same time span. Contrary to this model, Velikovsky's predictions about Venus and its cometary nature were as follows:

Venus is a hot planet because, until recently, it was a comet.

In the 1950s, the scientific consensus was that Venus was an old planet similar to Earth and Mars, and given that its orbit is also similar, its temperature should be as well. At the time, Venus' surface temperature was 'known'[322] to be -25°C,[323] and some scientists even believed that Venus might be habitable.[324]

But when Venus probe *Mariner 2* sent back its data in 1963, the scientific community was flabbergasted. Venus' average surface temperature was at least a whopping *425°C*.[325] The allegedly habitable planet has the temperature of molten lead and is the hottest planet in the solar system.[326]

The hot nature of Venus, besides its surface, was confirmed when researchers[327] measured gravitational variations over the planet, from which they deduced that Venus' crust was very thin, between 10 and 20 km,[328] compared to 'sister' planets like Earth or Mars whose crust thickness is between 50 and 100 km.[329]

Figure 77: © NASA
Planet Venus

This thin lithosphere indicates that Venus has *a hot, active interior* that prevents the crust from cooling down and hardening over a substantial thickness.

In conclusion, as predicted by Velikovsky, Venus is indeed a hot planet, on the surface as well

---

[322] Laird Scranton (2012), *The Velikovsky Heresies: Worlds in Collision and Ancient Catastrophes Revisited*, Simon and Schuster, p. 38.
[323] -13°F.
[324] Richard Lemmons (2017), "Goldilocks in space Earth Mars and Venus," *Climate Policy Watcher*.
[325] 864°F. NASA Content Administrator (2012), "Mariner 2 Data from Venus," *Nasa.com*.
[326] Fraser Cain (2013), "What is the Hottest Planet in the Solar System?" *Universe Today*.
[327] W. S. Kiefer, and B. H. Hager (1991), "A mantle plume model for the equatorial highlands of Venus," *Journal of Geophysics Research* 96:20947–20966.
[328] 6 to 12 miles.
[329] 31 to 62 miles.

as inside. This strongly suggests that not long ago Venus was still a blazing hot comet and that it has not yet fully cooled down from its previous cometary state.

Craters on Venus

Since it was believed that Venus was an old planet,[330] it was also assumed that it had been exposed to asteroids for billions of years and was, therefore, likely to be marked with many craters.

But at the time, the surface of Venus could not be directly observed due to its very dense atmosphere. In the 1970s, the first Venusian probes allowed direct observations of Venus' surface[331] and revealed that Venus had *a surprisingly low number of craters.*[332]

These repeated observations suggest that, as predicted by Velikovsky, Venus is a young planet. Until recently it was still a comet, therefore not enough time in its 'planet life' has elapsed for it to be impacted a large number of times.

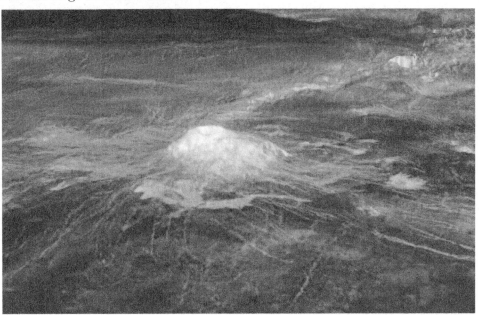

Figure 78: © NASA/JPL
The relatively pristine surface of Venus

Venus' Rotation

The young age of Venus is confirmed by analysis of argon isotopes. Argon-36 is abundant in young atmospheres, but because of its decay rate and volatility, it is almost absent in the atmospheres

---

[330] Scranton (2012), p. 32.
[331] Don Mitchell (2003), "First Pictures of the Surface of Venus," *Mentallandscape.com*.
[332] There are about 1,000 craters on Venus. For comparison Mars, despite its smaller size, is covered by 635,000 craters at least 0.6 km (1 mi) wide. See: Fraser Cain (2008), "Craters on Venus," *Universe Today*, and Space.com Editors (2012), "Mars Surface Scarred by 635,000 Big Impact Craters," *Space.com*.

of old planets. Concentration of Argon-36 in Venus is hundreds of time higher than on Earth.[333]

Conversely, Argon-40 is a decay product of Potassium-40, so its concentration should be high in old planets and low in young ones. It appears that Argon-40 concentration on Venus is fifteen times lower than on Earth,[334] confirming that Venus is much younger than Earth.

According to Velikovsky, because of its recent cometary nature and its chaotic interaction with Mars and Earth, Venus should display an anomalous rotation compared to other planets in the solar system.

This prediction was, like the others, considered heresy.[335] But in 1962, the U.S. Naval Research Laboratory in Washington announced that Venus had *an unexpected slow retrograde rotation*.[336] It is the only planet in the *inner* solar system to display a retrograde rotation.[337]

Confirming Venus' peculiar celestial movements, astronomers demonstrated that Venus' spin was in resonance with Earth's orbit[338] – each time Venus passes between the Sun and Earth, it shows the same side to Earth.

Such a resonance suggests a relatively recent close approach between Earth and Venus, which 'locked' the spin of the smaller planet (Venus) with the orbit of the larger one (Earth). In addition, one of the main arguments that sought to refute Velikovsky's theory was that Keplerian orbits cannot cross each other – so that collisions or near-collisions can't occur.

However, a groundbreaking paper published in 1974 demonstrated that:

- Keplerian orbits can cross each other,
- Venus could have had a highly elliptical (cometary) orbit in the recent past,
- the solar system could have exhibited stable planetary orbits before the arrival of Venus,
- Venus could have acquired a circular orbit soon after its integration to the solar system.[339]

Figure 79: © Imgur
Resonance pattern of Venus relative to Earth

Venus' Electrical Activity

Because of Venus' cometary nature and its past interactions with Mars, Velikovsky predicted that Venus should display some electrical activity. In the 1950s, this prediction was contrary to the scientific consensus, which considered Venus an electrically inert planet.

---

[333] John Noble Wilford (1978), "Argon Level on Venus Stirs Debate," *New York Times*.
[334] Billy P. Glass (1982), *Introduction to Planetary Geology*, Cambridge University Press.
[335] Scraton (2012), p. 39.
[336] Andrew J. Butrica (1996), "To See the Unseen: A History of Planetary Radar Astronomy," *NASA SP-4218*, chapter 2.
[337] Uranus is the only other planet having a retrograde rotation but it is located in the *outer* solar system.
[338] Peter Goldreich *et al.* (1966), "Spin-orbit coupling in the solar system," *Astronomical Journal* 71:425.
[339] L.E. Rose *et al.* (1974), "Velikovsky and the sequence of planetary orbit," *PENSEE Journal VIII*, Mikamar Publishing.

This view prevailed for a few decades. But with the advent of Venusian probes, the electric activity of Venus was suspected as early as the late 1970's.[340] Since then a growing body of evidence strongly suggests that lightning does occur on Venus.[341]

But lightning is only one of the manifestations of Venus' electrical nature. In 1991, plasma waves were observed in the Venusian atmosphere by space probe *Galileo*.[342] Along the same line, since the late 1960's the presence of an ionosphere[343] around Venus was suspected[344] and has since been ascertained.[345]

In 2013, the European Space Agency announced that Venus did not have a normal spherical ionosphere but a *teardrop-shaped ionosphere, i.e. a comet's tail*,[346] as depicted by the image below:

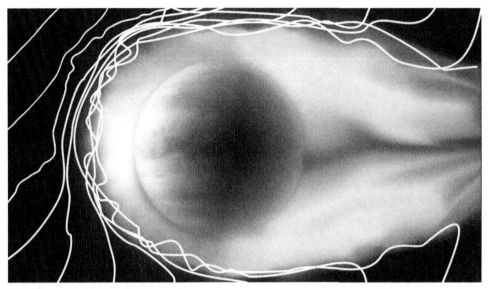

Figure 80: © ESA
Venus' tear-shaped ionosphere

Notice that Venus' comet tail is very long: 45 million km.[347] So long, in fact, that *its ion tail touches the Earth* when the Sun, Venus and Earth are aligned.[348]

---

[340] W. Taylor *et al.* (1979), "Evidence for lighting on Venus," *Nature* 279:614–616.

[341] J.M. Grebowsky *et al.* (1997), "Evidence for Venus lightning," *Venus II*, pp. 125–157. S.A. Hansell *et al.* (1995), "Optical detection of lightning on Venus," *Icarus* 117:345–351.

[342] D.A. Gurnett *et al.* (1991), "Lightning and plasma wave observations from the Galileo flyby of Venus." *Science* 253(5027):1522–1525

[343] Electrically charged layer of a planet's atmosphere.

[344] Mariner Stanford Group (1967), "Venus: ionosphere and atmosphere as measured by dual-frequency radio occultation of Mariner V," *Science* 158:1678–1683.

[345] M. Pätzold *et al.* (2007), "The structure of Venus' middle atmosphere and ionosphere," *Nature* 450:657–660.

[346] Miriam Kramer (2013), "Venus Can Have 'Comet-Like' Atmosphere," *Space.com*.

[347] 29 million miles.

[348] Jeff Hecht (1997), "Science: Planet's tail of the unexpected," *New Scientist* 2084.

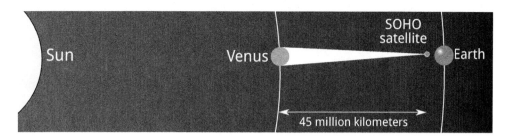

Figure 81: © New Scientist
Venus' ion tail

Notice also that Venus as a comet is not an isolated case in our solar system. In my previous book,[349] I described in detail how several planets of our solar system have acquired a number of new moons that were previously comets/asteroids.[350] The table below illustrates the increase in recently acquired moons[351]:

Planet \ Year	1975	2019
Jupiter	9	79
Saturn	10	82
Uranus	5	27

Figure 82: © Sott.net
Number of moons (1975 vs 2019)

---

[349] Lescaudron (2014) chapter 21.
[350] The difference between a comet and an asteroid is the electric discharge the body is subjected to. Unlike asteroids, comets are subjected to intense electric discharges, hence their glowing nucleus, coma and tail. See: Lescaudron (2014), chapter 18.
[351] Wikipedia Editors (2020), "List of Natural satellites," *Wikipedia*.

When Did the Water Transfer Occur?

The second part of this book started with a question about a major anomaly: during the Younger Dryas, a period of dramatic cooling, sea levels rose markedly instead of dropping.

Our hypothesis is that a massive dump of Martian ice is the cause of this anomaly; therefore the hypothesized close encounter with Mars should have happened *soon after the beginning of the Younger Dryas*, which is dated to 12,900 BP.

Notice that a Mars–Earth close encounter soon after 12,900 BP is compatible with the ALH84001 meteorite data that originated from Valles Marineris and whose estimated landing is 13,000 BP ±1,000 years.[352] It is also compatible with the Martian nakhlite meteorites that originated from Mars' oceanic coast and are believed to have fallen to Earth about 10,000 years ago.[353]

Now, are there other pieces of evidence that confirm this sequence of events? Can we clarify the time that elapsed between the beginning of the Younger Dryas triggered by cometary bombardments as depicted in Part I, and the Mars encounter with its ice dump?

As we will see below, several sources of information – among them ancient maps, reconstruction of past sea levels, past temperatures and moraine analysis – can give us a pretty clear idea of when the water transfer from Mars to Earth likely occurred.

Ancient Maps

Numerous maps dating back to the Renaissance show *an ice-free Antarctica*. For our purpose, we will only focus on two documents: the Oronteus Finaeus[354] map and the Buache maps,[355] but there are a number of others[356] that depict an unglaciated Antarctic continent, among which are the maps discovered by Robert Thome (1527), Sebastian Munster (1545), Giacomo Castaldi (1546), Abraham Ortelius (1570), Plancius (1592), Hondius (1602), Sanson (1651), Seller (1670) and, of course, the famous Piri Reis map,[357] which is shown below:

---

[352] A.J. Jull *et al.* (1989), "Trends in Carbon-14 Terrestrial Ages of Antarctic Meteorites from Different Sites," *Lunar Planetary Science* 20:488.
[353] See chapter "Evidence of Material Transfer from Mars to Earth."
[354] Oronce Finé (1494–1555), also known by his Latin name as Oronteus Finaeus, was a French mathematician and cartographer. See Charles Hapgood (1966), *Maps of the Ancient Sea Kings: Evidence of Advanced Civilization in the Ice Age*, Adventures Unlimited Press, pp. 79–113.
[355] Maps compiled in 1737 by French geographer Philippe Buache (1700–1773). Ibid., pp. 73–79.
[356] Ibid., p. 72.
[357] World map compiled in 1513 by the Ottoman admiral and cartographer Piri Reis (1465/70–1553). There are 207 fine charts drawn by Piri Reis in his map book containing correct and scientific information profusely presented. Piri Reis compiled a world map from his ancient sources in 1531. See ibid., p. 216.

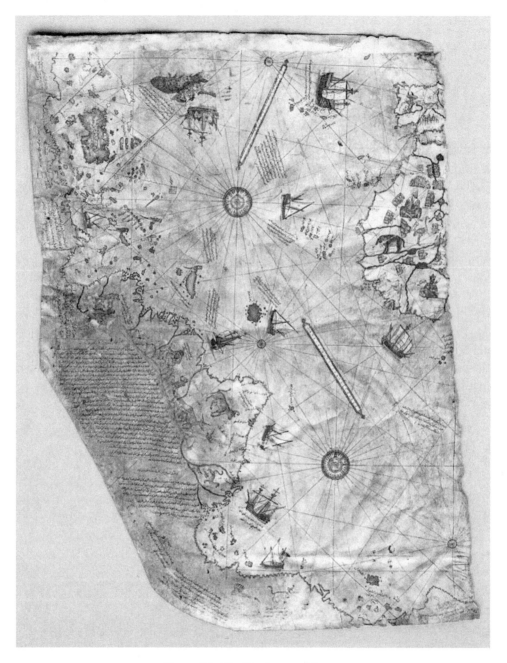

Figure 83: The Piri Reis world map

Over the years, these maps has been subjected to a number of thorough tests which revealed that the people who drew them had an excellent knowledge of longitudes, latitudes and spherical trigonometry,[358] a branch of geometry that reached its complete form only at the end of the 19th century.[359]

---

[358]Ibid., pp. 181–193.
[359]Solo Hermelin (2014), "Spherical Trigonometry – Update 12.11.12," *Solohermelin.com*.

It is also clear that the original designers of these maps had explored and surveyed the whole world and knew the exact size and circumference of our planet.

Of course, at first the maps were considered forgeries. After all, how could Renaissance geographers have knowledge that was so far ahead of their time? But, when scientists started to compare those ancient maps, including the ones depicting Antarctica, with modern maps, they found that the ancient maps were too accurate to be the result of random chance:

> In several years of research, the projection of this ancient map [Oronce Finé's map] was worked out. It was found to have been drawn on a sophisticated map projection, with the use of spherical trigonometry, and to be so scientific that **over fifty locations on the Antarctic continent have been found to be located on it with an accuracy that was not attained by modern cartographic science until the nineteenth century.**[360]

While these ancient maps were discovered between the 16th century and the 18th century, *Antarctica was only discovered at least one century later in 1820* and was ice-bound.[361] This suggests that these two maps are medieval copies of ancient originals drawn at a previous time when Antarctica was indeed ice-free. Also notice that the Buache map below shows an ice-free Antarctica comprised of two main islands.

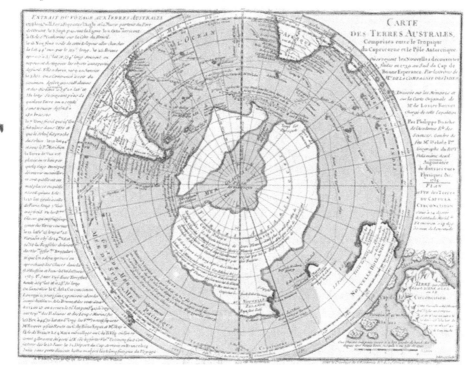

Figure 84: The Buache map of Antarctica (1737)

---

[360] Hapgood (1970), p. 258.

[361] Terence Armstrong (1971), "Bellingshausen and the discovery of Antarctica," *Polar Record* 15(99):887–889. While the first sighting of mainland Antarctica occurred in 1820, the first documented landing on Antarctica happened 33 years later, in 1853.

It was only during the 20th century that radar surveys enabled the mapping of the Antarctic bedrock[362] and confirmed that Antarctica is not, as previously believed, one solid single island, but rather an archipelago comprising two main islands as shown in the Buache map.

Similarly Orontius Finaeus' ancient map of Antarctica represents an ice-free continent. The maps were published in 1531 but were far more ancient than that. Apparently they were drawn by some very ancient people, then preserved for centuries and finally discovered and/or copied by Finaeus.[363]

Figure 85: © Christianus Wechelus
Ancient Orontius Finaeus map showing ice-free Antarctica

Not only does the Orontius Finaeus map depict Antarctica devoid of an ice sheet, it also reveals detailed rivers and estuaries, particularly around the Ross Sea, as shown on the top right of the map (see figure 85).

Modern science only discovered evidence of this feature in 2013, when sediment analysis from the coastal bottom of the Ross Sea revealed that they were river sediments that originated from an ice-free continent at least 6,000 years ago.[364]

A further examination of the Orontius Finaeus map reveals a number of other river inlets and islands along the coast of Antarctica, *which are now under water*. This suggests that at the time the original Orontius Finaeus map was drawn, the sea level was noticeably lower than it is today.

---

[362] P.A. Shumskiy (1959), "Is Antarctica a Continent or an Archipelago?" *Journal of Glaciology* 3(26):455–457.

[363] More will be revealed about ancient maps later in part 2, chapter "When did the Ice transfer occur." That's why the topic of ancient maps is covered so succinctly here.

[364] Hall (2013).

Some of the features are now more than 120 meters under water. As shown in the sea level diagram below, the only time over the last 135,000 years when water level was that low was *about 15,000 years ago*:

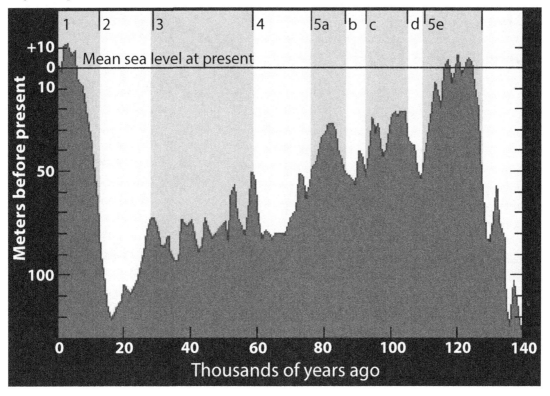

Figure 86: © NOAA
Sea level over the past 140 kY

Were the originals of these maps drawn about 15,000 years ago, when the water level was low enough for the now-submerged land features to be exposed? If these ancient maps representing an ice-free Antarctica *were* drawn about 15,000 years ago, then the close encounter with Mars, its massive ice dump and the subsequent rise in sea level must have happened later.

The previously mentioned sediment analysis from an ice-free Antarctica indicates that the original maps have to date to at least 6,000 years ago.

So we have a date range for the Martian ice dump *somewhere between 6,000 and 15,000 years ago*. Now, let's try to narrow this range.

Sea Level and Temperatures

The sea level drop (about 30 meters[365]) that should have been induced by the Younger Dryas cooling (12,900 to 11,700 years ago) didn't happen. Above we presented evidence suggesting that the sea level drop was offset by the intake of Martian water.

---

[365] 100 feet.

To know more precisely when this intake might have occurred, we need to find detailed sea level analysis and take a close look at it. Coral reef analyses provide such detailed sea level data:

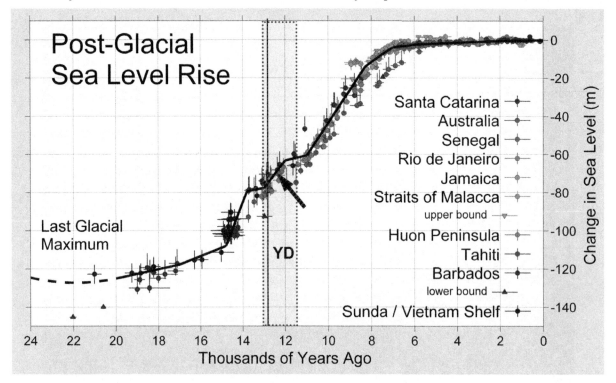

Figure 87: © Robert A. Rohde
Average change in sea level deduced from coral reefs

In the above diagram, the Younger Dryas period is represented by the light grey rectangle. Within this rectangle, we can see a *sudden rise* in sea level after centuries of stagnation, as indicated by the black arrow. It started a few centuries after the beginning of the Younger Dryas, ca. 12,600 BP, as shown by the black vertical line.

The hypothesized Martian ice dump was so massive – with a volume equivalent to the whole Arctic Ocean[366] – that it should have noticeably decreased the temperatures all over our planet. If this ice dump happened ca. 12,600 BP, ice cores should show a marked temperature drop around this date.

That's exactly what the reconstructed temperature history based on the analysis of oxygen-18 isotope reveals in the graph on the next page.

The diagram above shows that the beginning of the Younger Dryas is marked first by a drastic cooling, shown by the dashed vertical line, followed a few centuries later by a second abrupt cooling, depicted by the black vertical line.

Those two consecutive temperature drops suggest that the Younger Dryas was not triggered by one but *two consecutive major cooling events.*

---

[366]The volume of the Martian Ocean is estimated to 20,000,000 km^3 (about 5,000,000 cubic miles). See chapter "Water on Mars."

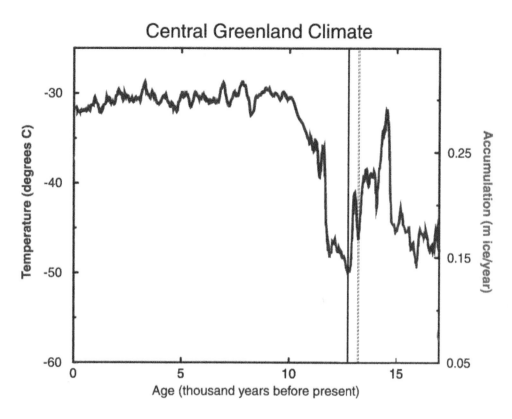

Figure 88: © Keit Alverson
Temperatures during the Younger Dryas

Moraine Analysis

The quick succession of two major cooling events at the beginning of the Younger Dryas is also indicated by moraine[367] analysis, as described in the following excerpt:

> **The Younger Dryas was not just a single climatic event.** Late Pleistocene climatic warming and cooling not only occurred before and after the YD, but **also within it.** All three major **Pleistocene ice sheets, the Scandinavian, Laurentide, and Cordilleran, experienced double moraine-building episodes, as did a large number of alpine glaciers.**
>
> Multiple YD moraines of the Scandinavian Ice Sheet have long been documented and a vast literature exists. The Scandinavian Ice Sheet **readvanced during the YD and built two extensive end moraines** across southern Finland, the central Swedish moraines, and the Ra moraines of southwestern Norway (Fig. 4). ^{14}C dates indicate **they were separated by about 500 years.**[368]

The graph below shows the double moraine found in Southern Finland and the marked advance of the ice extent. The substantial gap between the first and the second moraine, about 30 km,[369] is a testimony to the severity of the second cooling event:

---

[367] Moraines are geological formations that mark the limit of ice extent.
[368] Don J. Easterbrook *et al.* (2011), "Evidence for Synchronous Global Climatic Events," in *Evidence-Based Climate Science*, pp. 53–88.
[369] 18 miles.

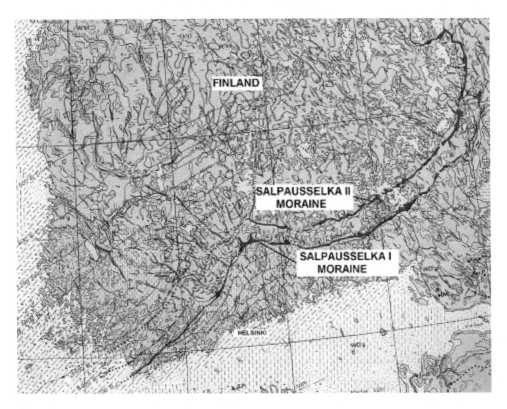

Figure 89: © Don J. Easterbrook *et al.*
Double Younger Dryas moraine of the Scandinavian ice sheet

Beyond the case of the Scandinavian ice sheet, Loch Lomond in Scotland also reveals multiple Younger Dryas moraines. The dating of these moraines is consistent with the two Scandinavian moraines separated by about 500 years:

> Among the first multiple YD moraines to be recognized were the Loch Lomond moraines of the Scottish Highlands. ... The Loch Lomond moraines consist of multiple moraines. **Radiocarbon dates constrain the age of the Loch Lomond moraines between 12.9k and 11.5k calendar years ago.**[370]

Ancient maps, temperature and sea level reconstruction, and moraine analysis provide a consistent picture, where the beginning of the Younger Dryas seems to have been marked by two distinct catastrophic cooling events that happened in close succession:

1. ca.12,900 BP: A major cometary bombardment, as described in the first part of this book and accepted by a growing number of scientists.

2. ca.12,600 BP: A few centuries later, a close encounter between Earth and Mars and the accompanying ice dump centered on Antarctica.

---

[370] Anthony Watts (2012)., "The Intriguing Problem Of The Younger Dryas—What Does It Mean And What Caused It?" *Wattsupwiththat.com*.

Conclusion

Figure 90: Nicolas Poussin - The Universal Deluge

The information gathered above enables us to hypothesize a scenario for the second event ca. 12,600 BP that involves the following steps:

1. Venus, a cometary body, enters the solar system and follows a typical eccentric cometary orbit around the Sun and Jupiter.

2. Comet Venus passes by Mars and disrupts its orbit.

3. Mars' disrupted orbit brings it very close to Earth.

4. The close proximity between Mars and Earth triggers a massive electric discharge, transferring Martian material, including its ocean, in the form of ice, to Earth.

The very short time span between the two events, about three centuries (a blink of an eye in the celestial timescale), makes us wonder if they are not somehow related. Maybe the comet Venus was part of a cometary swarm of which it constituted the main object. After entering the solar system, the Venus cometary swarm followed a typical Jupiter–Sun orbit.

This eccentric orbit passed close to Earth orbit and, during the first crossing, some of the bodies included in the swarm were attracted by Earth's gravity and provoked a substantial cometary bombardment with no less than five major meteors with diameters in excess of 10 km reaching Earth. This could be the catastrophic event (ca. 12,900 BP) that initiated the Younger Dryas. Because of its higher momentum, Venus then pursued its orbit around the Sun and Jupiter. About three centuries later, Venus came close enough to Mars to knock it out of its orbit and pushed it dangerously close to Earth, leading to the electric discharge described above.

This scenario is close to the one proposed by Velikovsky 70 years ago. The only substantial differences are the water transfer and, of course, the dating. In fact, the dating was the main argument brought against Velikovsky, who suggested the Earth–Mars encounter to have occurred between 3,500 and 2,800 BP. It is still the main bone of contention today, as illustrated by this excerpt summarizing the main weakness identified by the scientific community about Velikovsky's theory:

> So far, the only piece of the geologic evidence which has shown to have a catastrophic origin is a "raised beach" containing coral-bearing conglomerates found at an elevation of 1,200 feet above sea level within the Hawaiian Islands.
>
> The sediments, which were misidentified as a "raised beach", are now attributed to megatsunamis generated by massive landslides created by the periodic collapse of the sides of the islands.
>
> In addition, **these conglomerates, as many of the items cited as evidence for his ideas in Earth in Upheaval, are far too old to be used as valid evidence** supporting the hypothesis presented in Worlds in Collision.[371]

Indeed, the dating suggested by Velikovsky isn't backed up by much evidence in the form of major catastrophes affecting the whole planet, although there is a good case for a localized catastrophe in the Middle East during the Bronze Age, ca. 3,700 BP.[372]

On the other hand, the onset of the Younger Dryas (ca. 12,900 BP and 12,600 BP) offers plenty of evidence, as described in this book, for sudden and major changes over the entire planet.

Velikovsky considered that the second event, a close encounter between Mars and Earth and its accompanying water/ice dump, was referenced in mythology as the Great Flood. He based his dating mostly on the wrong chronology offered by the Old Testament, hence the ca. 2,800 BP dating. But the Hebrew mythology as recorded in the Old Testament is only one of numerous mythologies mentioning the Great Flood.

In 500 cultures spanning all inhabited continents, researcher Douglas Ettinger found that *about 90% included an account of a 'great deluge,'* as shown in the table below, where the 'deluge/flood'

---

[371] Wikipedia Editors (2006), "*Worlds In Collision,*" *Wikipedia*.

[372] Multiple lines of evidence suggest a Tunguska-like, cosmic airburst event that obliterated civilization – including the Middle Bronze Age (MBA) city-state anchored by Tall el-Hammam in the Middle Ghor (the 25 km diameter circular plain immediately north of the Dead Sea) ca. 3,700 BP. Analyses of samples taken over thirteen seasons have been performed with remarkable results. Commensurate with these results are the archaeological data collected from across the entire occupational footprint (36ha) of Tall el-Hammam, demonstrating a directionality pattern for the high-heat, explosive 3,700 BP Middle Ghor Event that, in an instant, devastated approximately 500km^2 immediately north of the Dead Sea, not only wiping out 100% of the MBA cities and towns, but also stripping agricultural soils from once-fertile fields and covering the eastern Middle Ghor with a super-heated brine of Dead Sea anhydride salts pushed over the landscape by the Event's frontal shockwaves. Based upon the archaeological evidence, it took at least 600 years to recover sufficiently from the soil destruction and contamination before civilization could again become established. See: Phillip Silvia *et al.* (2018), "The 3.7kaBP Middle Ghor Event: Catastrophic Termination of a Bronze Age Civilization," Conference: Annual Meeting of the American Schools of Oriental Research (ASOR).

column is in medium gray.[373] The prevalence of this myth in most cultures all across the planet suggests that the Deluge was truly a worldwide catastrophe.

As shown on the next page, the Old Testament is neither the only nor the oldest account of the Great Flood. For example, the OT is predated by the ancient Mesopotamian *Gilgamesh* epic,[374] which is about 4,000 years old.[375]

According to Alexander Heidel,[376] Mesopotamian and Hebrew myths could descend from an even older common original,[377] and in any case, the written version of the Epic was preceded by oral versions.

This suggests that there was indeed a Great Flood, but it happened much earlier than claimed by the Old Testament and Velikovsky, even earlier than the invention of writing ca. 5,200 BP.

In their mythology, most cultures have surprisingly similar descriptions of the Great Deluge[378] referring to a very real event that displays an uncanny resemblance with the event described above, which raises an obvious hypothesis: *the Great Flood refers to the close encounter between Mars and Earth 12,600 years ago which precipitated zillions of tons of ice and water on our planet.*

---

[373] Douglas Ettinger (2016), *The Great Deluge: Fact or Fiction?*, Ettinger Journals.
[374] Utnapishtim's Tale, tablet XI.
[375] Richard Hooker (2020), "General information on the Sumerian Epic Gilgamesh (ca. 2000 B.C.E.)," Arkansas State University.
[376] Assyriologist and biblical scholar (1907–1955).
[377] Alexander Heidel (1946), *The Gilgamesh Epic and the Old Testament Parallels*, University of Chicago.
[378] Michael Witzel (2012), *The Origins of the World's Mythologies*, Oxford University Press, pp. 177–180.

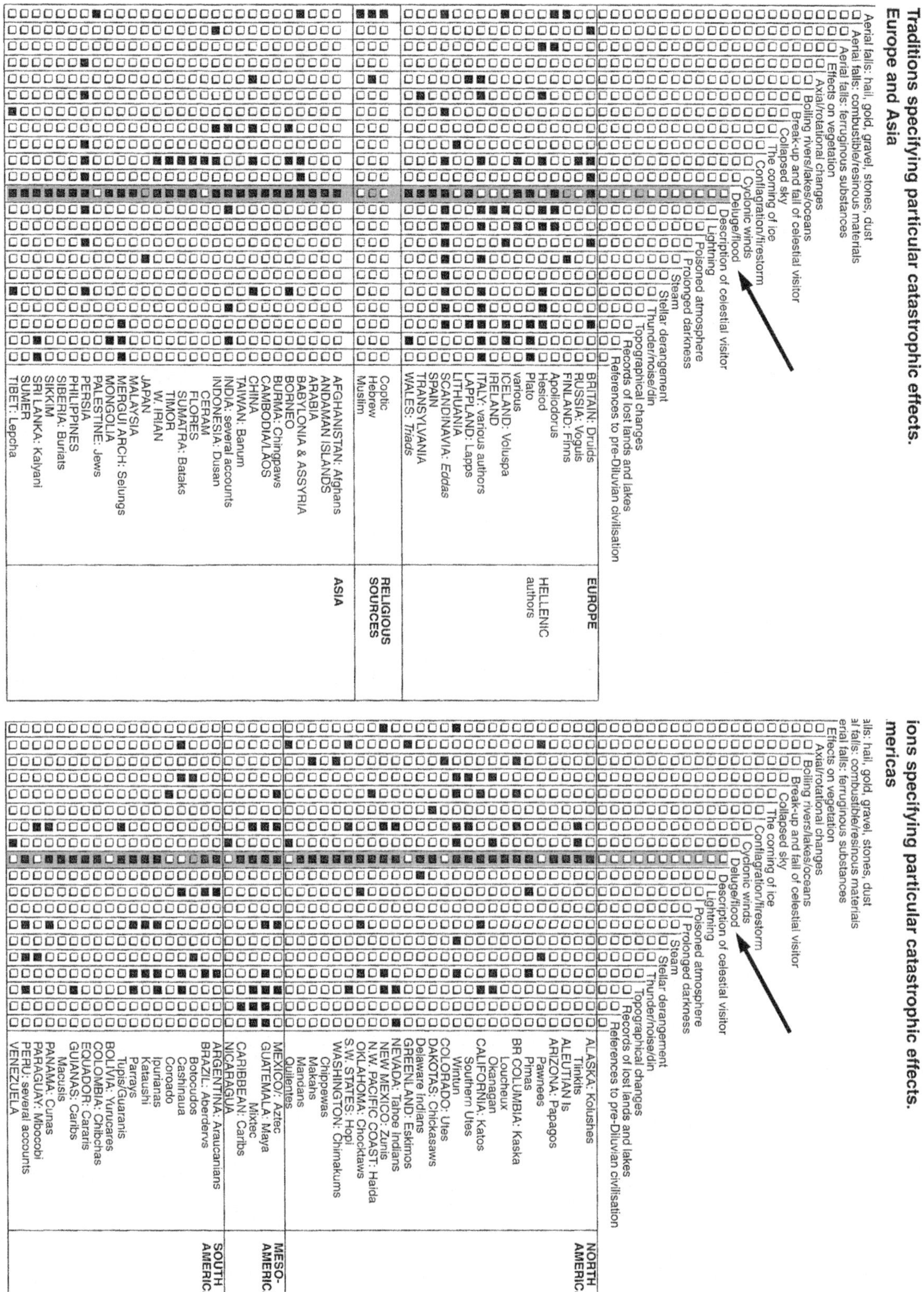

Figure 91: © Ettinger

Catastrophic features in traditional accounts (left: Eurasia, right: Americas)

Part III:
Volcanoes, Earthquakes and the 3,600-Year Comet Cycle

Figure 92: © Tomas van der Weijden
Erupting Klyuchevskaya Sopka volcano below a blazing meteor

In the previous two parts of this book, we proposed explanations for the events that triggered the Younger Dryas. In part I, we explained how, ca. 12,900 BP, several cometary fragments hit Earth and caused a marked global cooling. This time is indicated by the dashed vertical line in the diagram below.

In part II, we described how, about three centuries later, ca. 12,600 BP, an electric discharge transferred part of the water on Mars to Earth and drastically decreased the temperature of our planet, as shown by the black vertical line:

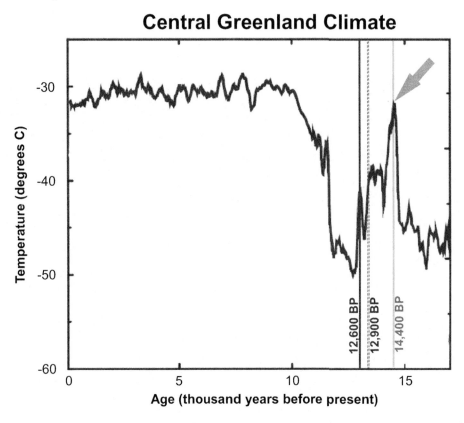

Figure 93: © Alley
Greenland temperature 18,000 BP – now

Now, while looking at this diagram, it is hard not to notice that the sudden cooling episodes mentioned above were only two of three catastrophic events that preceded the Younger Dryas. We can distinctly see that *a third cooling event occurred ca. 14,400 BP*, as shown by the medium grey vertical line and arrow.

This event had an even *greater* magnitude than the two events that followed, since it induced a 10°C[379] drop as compared to the two following events which 'only' induced a 7°C[380] drop.

In this third part, we will explore the specifics of the 14,400 BP event and explain how it might be part of a larger 3,600-year cometary cycle.

---

[379] 18°F.
[380] 12°F.

The 14,400 BP (12,400 BC) Event

The sudden temperature drop ca. 14,400 BP marks the beginning of the *Older* Dryas,[381] a cooling period that lasted a few centuries.

But the temperature reconstructions[382] shown in the previous diagram displays only one atmospheric parameter: temperature for a single location, Greenland.

So, let's have a look first at the geographic extent of the 14,400 BP temperature anomaly. The sediment cores from the Cariaco basin in Venezuela reveal a pattern similar to Greenland with a marked drop in temperature – about 3°C[383] – as shown by the vertical black line and the arrow in the diagram below:

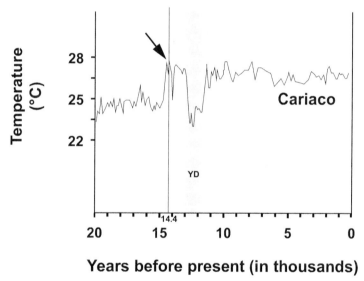

Figure 94: © NCDC NOAA
Cariaco temperatures from 20,000 BP to now

In the same vein, the glaciers in northern Scotland were even thicker and deeper during the Older Dryas than the succeeding Younger Dryas.[384] In Belgium, the analysis of multiple variables like temperature, fauna, flora and oxygen on the Moervaart paleolake sediment confirm the Older Dryas cooling:

---

[381] Jibin Xue *et al.* (2009), "A new high-resolution Late Glacial-Holocene climatic record from eastern Nanling Mountains in South China," *Chinese Geographical Science* 19:274–282.
[382] Based on oxygen-18 isotope.
[383] 5°F.
[384] Paul Pettit, and Mark White (2012), *The British Palaeolithic: Human Societies at the Edge of the Pleistocene World*, Abingdon, UK: Routledge, pp. 374, 477.

... drier and **colder climate** with more barren ground and a larger abundance of grasses in the vegetation, resulting in the development of a grass-steppe tundra. Based on the pollen biozones, these deposits appear to correlate with the Older Dryas.[385]

The reconstruction of sea surface temperature through the study of alkenones[386] in the Pacific Ocean sediment cores conducted in southern Okinawa, Japan, show a cooling starting ca. 14,300 BP:

> **A cooling stage at 14.3 to 13.7 kyr BP was observed** between Bølling and Allerød warm phases, which is coeval with so-called Older Dryas event.[387]

However, ice cores from Vostok, Antarctica, reveal a very minute temperature drop, as shown by the black arrow in the following diagram:

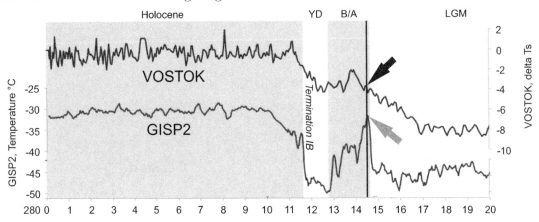

Figure 95: © Moffitt
Temperatures in Greenland and Vostok

This temperature decrease – indicated by the medium grey arrow above – is less marked than in Greenland, Venezuela, Belgium or Japan, which suggests that the 14,400 BP cooling event *affected the northern hemisphere and equatorial regions more than the southern hemisphere.*

Besides this temperature drop, ice core analysis reveals changes in several other atmospheric parameters, among them methane[388] (CH_4) and carbon dioxide[389] (CO_2). Both gases are indicators of biomass burning – forest fires, for example.

The diagram below shows this dramatic increase in methane, as indicated by the grey arrow, and the pronounced increase in carbon dioxide, as shown by the black arrow ca. 14,400 BP:

---

[385] J. A. Bos, *et al.* (2017), "Multiple oscillations during the Lateglacial as recorded in a multi-proxy, high-resolution record of the Moervaart palaeolake (NW Belgium)," *Quaternary Science Reviews* 162:26–41.

[386] Alkenones are unsaturated ketones, which represent the main organic biomarker proxy for sea surface temperature. It is based on the measurement of the relative abundance in marine sediments of unbranched long-chained methyl ketones with 37 carbon atoms.

[387] Jiaping Ruan *et al.* (2015), "A high resolution record of sea surface temperature in southern Okinawa Trough for the past 15,000 years," *Palaeogeography, Palaeoclimatology, Palaeoecology* 426:209–215.

[388] E. Dlugokencky *et al.* (2014), *Encyclopedia of Atmospheric Sciences*, Academic Press, chapter "Chemistry of the Atmosphere – Methane."

[389] F. Cherubini *et al.* (2011), "CO2 emissions from biomass combustion for bioenergy: atmospheric decay and contribution to global warming," *GCB Bioenergy* 3:413–426.

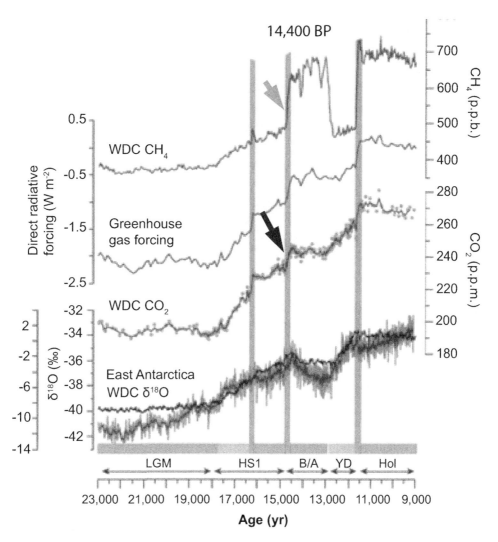

Figure 96: © Marcott
CH₄ and CO₂ concentration in Antarctica WAIS ice core (23 kA–9 kA)

Besides being biomass-burning byproducts, methane and carbon dioxide are also *two of the main gaseous components of cometary tails.*[390]

Sulfur[391] tells a similar tale, with an unusually high concentration in ice cores during the millennium from 15,000 BP to 14,000 BP.[392] In addition, as indicated by the black arrow and vertical line in the diagram below, a prominent sulfur spike appears ca. 14,400 BP. It is the seventh largest sulfur signal recorded over the past 25,000 years:

---

[390] Manuela Lippi (2010), "The composition of cometary ices as inferred from measured production rates of volatiles," *Carolo-Wilhelmina University*.
[391] In its sulfate ionic form: SO_4^{2-}.
[392] This millennium records 17 signals above 75 ppm sulfate. For comparison Krakatoa "only" released 40 ppm. See: G.A. Zielinski *et al.* (1996), "A 110,000-Yr Record of Explosive Volcanism from the GISP2 (Greenland) Ice Core," *Quaternary Research* 45(2):109–118.

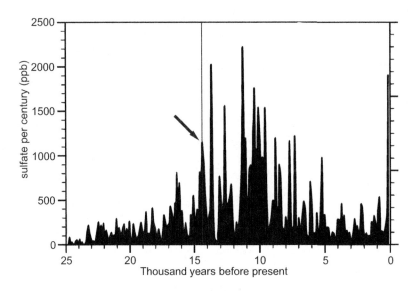

Figure 97: © Sott.net
Sulfur concentration 25,000 BP–now

The diagram below shows a concentration spike in calcium[393] found in Greenland ice cores ca. 14,400 BP, as indicated by the dark grey arrow and light grey vertical line:

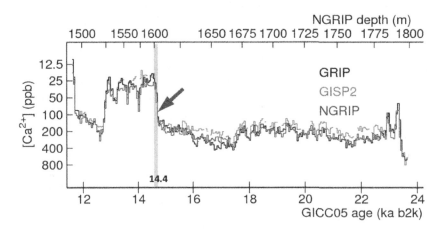

Figure 98: © Sott.net
Calcium (Ca) concentration in ice cores 24,000 to 11,000 BP

Those two concomitant spikes in sulfur and calcium concentrations are interesting markers because they can be caused by cometary impact ejecta:

Sulfur in the impactor or in sulfur-containing target rocks can be injected into the atmosphere in a vapor-rich impact plume. In some impact events, such as Chicxulub, the rocks hit by the impactor contain sulfur. Sedimentary rocks hit by an impactor sometimes include large amounts of evaporites. Evaporites are rocks that are formed with minerals that precipitated from evaporating water, such as halite (rock salt) and calcite (**calcium** carbonate). **Two other**

---

[393]In the Ca_2^+ ionic form.

very common evaporite minerals are gypsum ($CaSO_4 + H_2O$) and anhydrite ($CaSO_4$), both of which contain sulfur (S).[394]

To recap, the 14,400 BP event reveals sudden spikes in carbon dioxide, methane, sulfur and calcium, accompanied by a drastic temperature drop. These dramatic atmospheric changes expectedly had a major impact on life on Earth, as noted by Argentinian archaeologist Gustavo Politis:

> Most archeological sites with unquestionable extinct mega-fauna have Late Pleistocene dates from c. **14,400** to 13,000 BP.[395]

It appears that the 14,400 BP event was the starting point of a megafauna extinction that lasted more than a millennium and was worsened by the 12,900 and the 12,600 BP events.[396] The magnitude of this mass extinction was exceptional:

> The late Quaternary megafaunal extinctions at the end of the Pleistocene, resulting in the **loss of between 35 and 90% of large-bodied animal species** on ice-free continents (excluding Africa), represented the **most profound faunal transition that Earth's ecosystems experienced during the Cenozoic.**[397]

For reference, the Cenozoic age (meaning 'new life') spans from 65 million years ago, when the Chicxulub cometary impact ended the reign of the dinosaurs,[398] to now. Never during those 65 million years had a mass extinction been this dramatic. For example, the cave lion[399] and the woolly rhinoceros[400] were two representatives of the megafauna species that became extinct ca. 14,400 BP:

Figure 99: © Britannica.net
Artist representations of the woolly rhinoceros and the cave lion

---

[394] David Kring (2020), "Impact induced perturbations of atmospheric sulphur," *Lunar and Planetary Institute*.
[395] Politis Gustavo *et al.* (2019), "Campo Laborde: A Late Pleistocene giant ground sloth kill and butchering site in the Pampas," *Science Advances* 5(3).
[396] See description of the Younger Dryas mass extinction in part I, chapter "Younger Dryas."
[397] Politis (2015).
[398] Charles Choi (2013), "Asteroid Impact That Killed the Dinosaurs: New Evidence," *Live Science*.
[399] Anthony Stuart *et al.* (2011), "Extinction chronology of the cave lion Panthera spelaean," *Quaternary Science Reviews* 30:2329–2340.
[400] Edana Lord *et al.* (2020), "Pre-extinction Demographic Stability and Genomic Signatures of Adaptation in the Woolly Rhinoceros," *Current Biology*.

Now, it would be interesting to see if there is an impact crater that might explain the 14,400 BP event. Such a crater should exhibit the following characteristics:

- Time match, ca. 14,400 BP.

- Northern location, since the impact seems to have been more pronounced in the northern hemisphere as suggested by the Vostok and Greenland temperature diagrams shown above.

- Large impactor, given the magnitude of the effects.

Searching for impact craters on Earth is not an easy task because wind, rain, earthquakes, sea-level changes, vegetation growth and urbanization tend to erase these geological features.[401] Plus, the search for impact craters doesn't seem to be a priority for modern science, perhaps due to the strong ideological resistance – uniformitarianism – to the very idea of relatively recent cometary impacts and their obvious catastrophic consequences.

Nonetheless, at least three databases, EDEIS,[402] Somerikko[403] and EID,[404] list some of the impact craters found on Earth. As of 2019, there are approximately 200 confirmed impact craters and 700 probable/possible ones.[405]

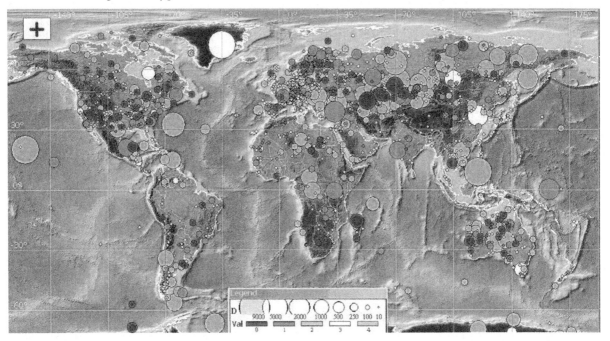

Figure 100: © Gusiakov
The 905 impacts found on the surface of Earth divided by their age of formation

---

[401] Lunar and Planetary Institute Editors (2020), "Shaping the Planets: Impact Cratering," *Lunar and Planetary Institute*.

[402] "Expert Database on Earth Impact Structures," compiled by the Institute of Computational Mathematics and Mathematical Geophysics SD RAS Tsunami Laboratory in 2008.

[403] Jarmo Mollanen (2009), "Impact structures of the World," *Somerikko.net*.

[404] Carlyle Beals (1955), "Earth Impact Database," *Dominion Observatory*.

[405] David Rajmon (2019), "Suspected Earth Impact Sites database," *David.rajmon.cz*.

The lack of investigation in this domain also affects the dating of the craters. Up to now, *most craters have not been dated*, and the ones that have been usually come with a broad impact time range. As a result, there are a number of large craters located in the northern hemisphere whose impact time range encompasses the 14,400 BP date. One of them was recently discovered by professor in Quaternary science Kurt Kjaer[406] and is located beneath Hiawatha glacier in northwest Greenland. This is a 31-kilometer-wide,[407] circular bedrock depression created by an iron-rich meteorite whose diameter is estimated to have been 1.5 km.[408]

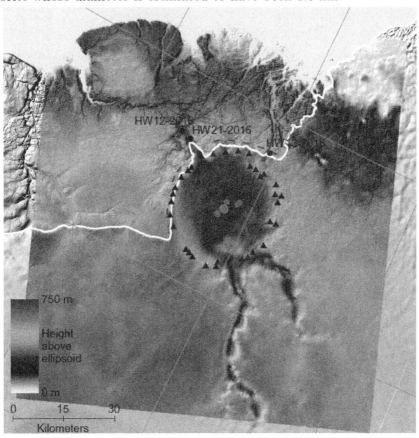

Figure 101: © Kjaer
Topography of the Hiawatha crater

The circular shape of the crater suggests that the meteorite was not part of the 12,900 BP swarm that flash-froze the woolly mammoths, because the trajectory of this swarm was almost tangential with the North Pole, hence the elliptical craters.[409]

Another reason for considering an ice sheet impact, including Greenland ca. 14,400 BP, is that it would explain a scientific anomaly known as Meltwater Pulse 1A (MWP1A).[410]

---

[406] Kurt Kjær (2018), "A large impact crater beneath Hiawatha Glacier in northwest Greenland," *Science Advances*.
[407] 20 miles.
[408] 1 mile.
[409] For more details about this point see Part I, chapter "Atmospheric ablation induced by a cometary impact."
[410] T.M. Cronin (2012), "Rapid sea-level rise," *Quaternary Science Reviews* 56:11–30.

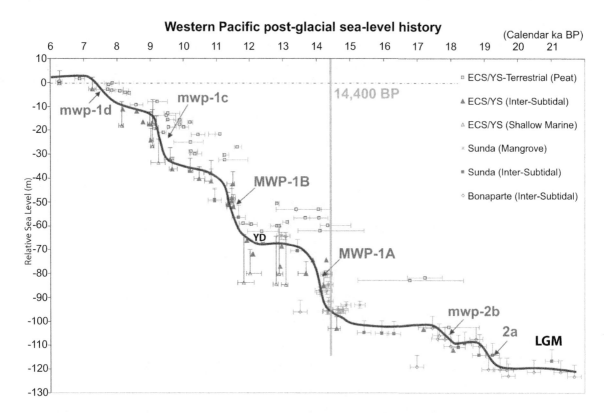

Figure 102: © Liu
Sea-level history (22,000 BP–6,000 BP)

The diagram above was made from coral analysis conducted by oceanographer Jean Paul Liu.[411] It shows that MWP-1A started ca. 14,400 BP, as indicated by the grey vertical line.

MWP-1A was a massive water release leading to a 20-meter rise[412] in sea level over a few centuries. The oddity is due to the fact that the *melting* occurred during the above-mentioned Older Dryas, which was a pronounced *cooling* period, usually accompanied by a sea-level *drop*.

So, how can we reconcile *cooling* and *melting*? One solution would be a substantial cometary body impacting an ice sheet. It would lead to the *melting* of the ice sheet, and also to global *cooling* due to the large quantity of ejecta[413] (dust, ice crystals) released in the atmosphere stimulating the nucleation of clouds and reducing the penetration of sun rays.[414]

In any case, while it is highly probable that the observed disturbances ca. 14,4000 BP – atmosphere, extinction, meltwater – were related to a cometary event, it is still highly speculative at this point to designate the Hiawatha crater as the culprit. More data is needed about this crater, which was only discovered in 2016 and is buried under 2 km of ice.[415]

---

[411] Jean Paul Liu *et al.* (2004), "Reconsidering melt-water pulses 1A and 1B: Global impacts of rapid sea-level rise," *Journal of Ocean University of China* 3:183–190.

[412] Cronin (2012).

[413] K. Kaiho *et al.* (2017), "Site of asteroid impact changed the history of life on Earth: the low probability of mass extinction," *Scientific Reports* 7(14855).

[414] For more details on how cometary impacts induce global cooling, see Part I, chapter "The Tragic Fate of the Woolly Mammoths."

[415] Kjaer (2016).

Cometary Cycle?

The previous chapter shows that a catastrophic event, probably of a cometary nature, happened ca. 14,400 BP. This number piqued my interest because it is a multiple of 3,600 (4 x 3,600 = 14,400).

You may remember that Zecharia Sitchin provoked a lot of media attention and controversy with the release of his first book, *The 12th Planet,* in 1976.[416]

Based on his interpretation of Sumerian iconography, Sitchin postulated that a planet called *Nibiru* – the 12th planet of our solar system – followed a long, elliptical orbit, reaching the inner solar system every 3,600 years.

This hypothesis doesn't make much sense, because an astronomic body exhibiting a very elongated orbit and crossing our solar system can't be a (dim) planet. But it might very well match a (bright) cometary orbit.

In the drawing to the right, we can see a cometary trajectory, as represented by the dotted light grey curve, which passes through different electric field lines, illustrated by the concentric circles numbered +1, +2, +3 and +4 centered around the Sun.

An electric field line defines locations where the electric potential is the same. It's similar to the altitude lines on a geographic map where every point of the line is at the same altitude.

These changes in electric potential difference between the comet and its surrounding space trigger intense current, including electric discharges, between the comet and its surrounding space, leading to an overheated and glowing cometary body. That's why *an astronomical body following a very elliptical orbit around the Sun with a 3,600-year period cannot be a (dim) planet but has to be a (bright) comet.*

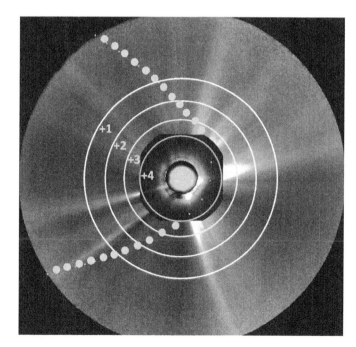

Figure 103: © Sott.net
Solar electric field and cometary orbit

Conversely, the electric potential at a given distance from the Sun being roughly the same, the astronomic bodies following a circular or slightly elliptical orbit will go through space exhibiting

---

[416] Zecharia Sitchin (1976), *The 12th Planet*, Simon and Schuster.

a constant electric potential. Therefore there is a balance between the electric potential of the body and the surrounding space. In this case, no electrical discharge occurs and the astronomical body doesn't glow.

In this sense, the fundamental difference between a comet and planet is not a matter of composition but a matter of *electrical activity*, which is related, among other factors, to the eccentricity of the orbit.

Thus, *a planet-sized comet is simply a glowing planet and a planet is a non-glowing comet.* Consequently the very same body can, successively, be a comet, then a planet, then a comet again, etc., depending on the variations in the ambient electric field to which it is subjected.

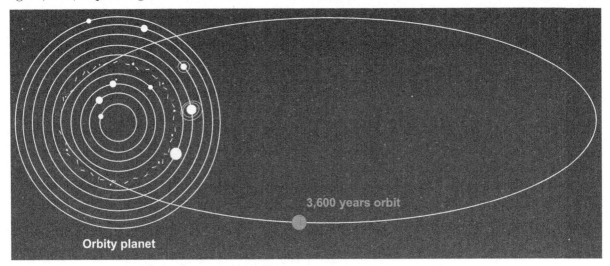

Figure 104: © Sott.net
3,600-year solar orbit

The unlikeliness of a planet exhibiting a very eccentric orbit aside, Sitchin's work was a good incentive to look further into Sumerian astronomy. Two features are particularly noteworthy:

First, the Sumerians had an excellent knowledge of astronomy in general and comets in particular. For example, Sumerian tablet *K8538*[417] describes with great accuracy the comet that struck Um-Al-Binni ca. 4,200 BP.[418] Actually *K8538* is the world's first scientific documentation on the approach and terrestrial impact of a large comet on Earth.

Second, the Sumerians had an elaborate sexagesimal[419] numeral system, which five millennia later is still used across the world for measuring parameters like angles, time or geographic coordinates. Within this numeral system, one of the main units was the *Sar*,[420] which is equal to 3,600 years.[421]

So, when combining the great Sumerian knowledge about comets and their 3,600-year time

---

[417] Frank Lemke *et al.* (2014), "The Sumerian K8538 tablet – The great meteor impact devastating Mesopotamia."
[418] The 4,200 BP event is one of the most severe climatic events of the Holocene. See: Peter De Menocal (2001), "Cultural Responses to Climate Change During the Late Holocene," *Science* 292(5517):667–673.
[419] Sexagesimal is a numeral system with sixty as its base.
[420] Also spelled "Shar" or "Saros."
[421] Christine Proust (2009), "Numerical and Metrological Graphemes: From Cuneiform to Transliteration," *Cuneiform Digital Library Journal*.

unit, one obvious question arises: is the *Sar* time unit just a coincidence or is it actually based on an astronomic constant, like a cometary cycle?

The idea of a 3,600-year periodic comet visiting our planet is usually dismissed because of the pull that would be exerted by stars other than the Sun on the comet. The reasoning usually goes as follows:

> Based on the only two bits of information we have about such an hypothetical comet which 1) has an orbital period of 3,600 years and 2) must pass within 1 AU of the Sun (because, if it doesn't, it can't fly by Earth), we can determine that this hypothetical comet must have an aphelion (the point in the orbit which is furthest from the Sun) of 469 AU (469 times the Earth–Sun distance).
>
> For comparison, this is 10x the distance between Pluto and the Sun. **The Sun's gravity is very weak at 469 AU (that is about 2.7 light days), so it would be easy for a passing object or another planet in our solar system to destabilize the comet's orbit and have it thrown into interstellar space.**

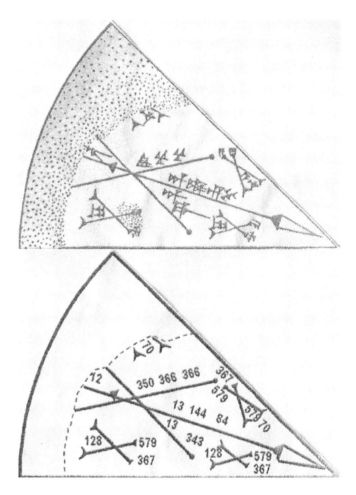

Figure 105: © Seifert
Section of the K8538 tablet.
Top: original cuneiform numbers.
Bottom: translation into Arabic numbers.

So is a 3,600-year stable orbit really impossible? Keep in mind the closest star to us is Proxima Centauri, about 4.25 light years away, so a comet with a 3,600-year orbit would remain within 2.7 light days from the Sun and at a far greater distance – at least 570 times greater – from other stars. Thus the comet would remain under the gravitational control of the Sun during its whole orbit, including its aphelion.[422]

Observations confirm this theory. Comets visiting the solar system and exhibiting a very long orbital period can follow a stable periodic orbit. This is, for example, the case with the great comet of 1811,[423] which exhibits an orbital period exceeding 3,000 years.[424]

---

[422] Furthest point from the Sun.
[423] C/1811 F1 also known as Napoleon's Comet.
[424] Gary Kronk (2020), "C/1811 F1 (Great Comet)," *Cometography.com*.

Figure 106: Depiction of the Great Comet of 1811

Theory and empirical observation show that a comet *can* exhibit a 3,600-year period. And since we have discovered previously that the 14,400 BP (3,600 x 4) event is probably related to a cometary event, let's pursue our investigation and focus on the hypothetical manifestation of the cometary cycle that is the closest to us, because as a general rule, the closer the date, the more data available.

In other words, did a cometary event occur ca. 3,600 (3,600 x 1) years ago?

The 3,600 BP (1,600 BC) Event

Similarly to 14,400 BP, 3,600 BP was marked by a sudden drop in temperature, as shown by the black vertical line and the black arrow in the ice core temperature diagram below:

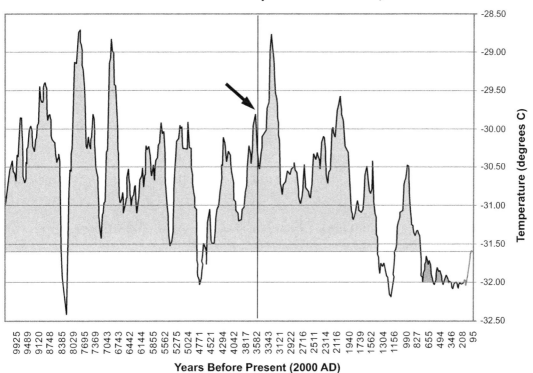

Figure 107: © Eastbrook
GISP2 temperature reconstruction over the past 10,000 years

Temperature as measured through O18 isotope is not the only atmospheric parameter disrupted ca. 3,600 BP. In the diagram below, we can see a 140-ppm spike in sulfate concentration that is about ten times larger than the 15-ppm background concentration exhibited during the following century:

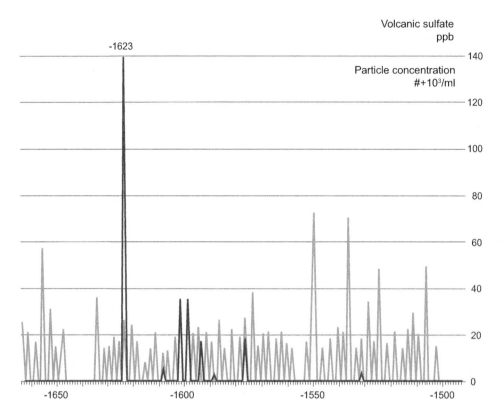

Figure 108: © Cybis.se
Sulfate spike in ice core ca. 1628 BC

The sulfate spike that happened about 3,600 years ago – 1628 BC – is concomitant with the eruption of Thera[425]:

> **A cataclysmic eruption of Thera (Santorini) was dated to 1628BC** from carbon dating of ash, and from tree rings as far away as Irish bogs, and Californian bristlecone pines. **The Thera explosion, maybe fifty times larger than Krakatoa** ...[426]

For reference, the eruption of Krakatoa, Indonesia, in 1883 was so violent that its explosions were heard in Mauritius, 4,800 km away.[427] Its pressure wave circled the globe three and a half times.[428] Ash was propelled 80 km high.[429] It killed tens of thousands of people and average temperatures fell by as much as 1.2 °C[430] in the year following the eruption.[431]

---

[425] Later known as Santorini, a volcano located in the Eastern Mediterranean Sea.
[426] Will Slatyer (2014), *Life/Death Rhythms of Ancient Empires – Climatic Cycles Influence Rule of Dynasties*, Partridge India, p. 8.
[427] 3,000 miles. The Independent Editors (2006), "How Krakatoa made the biggest bang," *The Independent*.
[428] Royal Society's Krakatoa Committee (1888) *The Eruption of Krakatoa: And Subsequent Phenomena*, Trübner & Company.
[429] 50 miles. NOAA Editors (2017), "On This Day: Historic Krakatau Eruption of 1883," *NOAA – National Centers for Environmental Information*.
[430] 2.2°F.
[431] Raymond Bradley (1988), "The explosive volcanic eruption signal in northern hemisphere continental temperature records," *Climatic Change* 12(3):221–243.

The eruption of Thera is estimated to have been *many times larger* than the Krakatoa eruption. But despite its exceptional magnitude, it falls short in explaining the major spike in atmospheric sulfate that occurred during the first half of the 17th century BC. Actually, Thera seems to have been a very small dust contributor:

> If temporal data alone cannot definitively link specific eruptions with distant climate records, other geochemical analyses must be used. For example, the ice-core acidity[432] layer contained high levels of sulfate but recent petrological calculations of the **sulfur emissions from Thera account for only 3-6% of the amount expected from the concentration of [sulfuric] acid in the ice layer.**[433]

If Thera contributed only about 5% of sulfur, then where did the remaining 95% come from?

Figure 109: © NASA
Remains of Thera caldera, which is about 18 km[434] in diameter.

It appears that Thera was not the only active volcano during this period of time:

> Minute shards of volcanic glass recovered from the 1645 ± 4 BC layer in the Greenland GRIP ice core have recently been claimed to originate from the Minoan eruption of Santorini [Hammer et al., 2003]. This is a significant claim because a precise age for the Minoan eruption provides an important time constraint on the evolution of civilizations in the Eastern Mediterranean.

---

[432] Sulfate atmospheric emissions combine with water to produce rainfall containing sulfuric acid, hence its detection in ice cores and the mention of "acidity" in the quoted paper.
[433] J.S. Vogel *et al.* (1990a), "Vesuvius/Avellino, one possible source of seventeenth century BC climatic disturbances," *Nature* 344:534–537.
[434] 11 miles.

There are however **significant differences between the concentrations of SiO2, TiO2, MgO, Ba, Sr, Nb and LREE**[435] between the ice core glass and the Minoan eruption, such that they cannot be correlatives. New chemical analyses of tephra from the Late Holocene eruption of the Aniakchak Volcano in Alaska, however, show a remarkable similarity to the ice core glass for all elements, and this eruption is proposed as the **most likely source of the glass in the GRIP ice core**. This provides a precise date of 1645 BC for the eruption of Aniakchak and is the first firm identification of Alaskan tephra in the Greenland ice cores.[436]

Dendrochronology analysis, based on Irish oaks and Swedish pines, confirms that the eruption of Aniakchak occurred in the first half of the 17th century. In addition, it often provides more reliable dating than ice cores. According to this dating, Aniakchak and Thera's eruption were virtually concomitant:

> Tree-ring data has shown that a large event interfering with normal tree growth in North America occurred during 1629–1628 (±65 years) BC. Evidence of a climatic event around 1628 BC has been found in studies of growth depression of European oaks in Ireland and of Scotch pines in Sweden. Bristlecone pine frost rings also indicate a date of 1627 BC, supporting the late 1600s BCE dating.
>
> However, McAneney and Baillie argue that there is a **chronological error in the Greenland ice core dates with ice core dates being around 14 years too old** in the 17th century BC, **thus implying that the eruption of Aniakchak, and not Thera, may have been the cause of the climatic upset evidenced by northern hemisphere tree-rings around 1627 BC**.[437]

Thera, in the Mediterranean Sea, and Aniakchak, in Alaska, were massive eruptions but apparently they are not the only volcanoes that went active around 1627 BC.

After the thorough analysis of tree rings and ice cores, geologist J.S. Vogel showed that the Avellino eruption was also partly responsible for the climatic disturbances observed in the beginning of the 17th century,[438] which witnessed a coincidence of *a number of eruptions*, including the Minoan eruption on Santorini.[439]

The above-mentioned Avellino eruption refers to an eruption of Mount Vesuvius, Italy. It is estimated to have had a VEI[440] of 6,[441] making it larger and more catastrophic than Vesuvius' more famous and well-documented eruption that wiped out Pompeii, Italy, in 79 AD.[442]

On top of that, Vogel discovered that Mount Saint Helens, USA, also went through a major eruption during the 17th century BC.[443]

---

[435] Light Rare Earth Elements. See: Eni Generalic (2019), "Rare Earth Elements," *Periodni.com*.

[436] N.J.G. Pearce *et al.* (2004), "Identification of Aniakchak (Alaska) tephra in Greenland ice core challenges the 1645 BC date for Minoan eruption of Santorini," *Geochemistry, Geophysics, Geosystems* 5(3).

[437] Jonny McAneney *et al.* (2019), "Absolute tree-ring dates for the Late Bronze Age eruptions of Aniakchak and Thera in light of a proposed revision of ice-core chronologies," *Antiquity* 93(367):99–112.

[438] J.S. Vogel *et al.* (1990b), "Letters to Nature: Vesuvius/Avellino, one possible source of seventeenth century BC climatic disturbances," *Nature* 344(6266):534–537.

[439] D.A. Hardy (1989), *Thera and the Aegean World, Volume III—Chronology*, Proceedings of the Third International Congress.

[440] Volcanic Explosivity Index.

[441] S.W. Manning (1992), "Thera, Sulphur and Climatic Anomalies," *Oxford Journal of Archaeology* 11:245–253.

[442] Andrew Wallace-Hadrill (2010), "Pompeii: Portents of Disaster," *BBC History*.

[443] Vogel (1990b).

To recap, we have at least four major and nearly simultaneous eruptions, namely Thera, Aniakchak, Vesuvius and St. Helens, that happened ca. 1628 BC, but this number of strangely synchronous major eruptions still doesn't account for the full sulfur spike found in various ice cores:

> If aerosols from Avellino approximated that of the very similar AD 79 eruption, they would have been predominately sulphurous and **three times** more abundant than those from the larger Theran eruption. Mount St Helens Yn may have yielded only **half** as much sulphate as Thera, but Aniakchak II could have produced sulphur emissions **four times** greater than Thera's.[444]

From the ratios suggested above by Vogel, we can deduce the following:

- Thera contributed about 5% of the total sulfur emission.

- Vesuvius (Avellino) contributed 15%.

- St. Helens contributed 2.5%.

- Aniakchak contributed 20%.

So, those four eruptions, despite their exceptional magnitude, *only account for about 42% of the total atmospheric sulfur* found in ice cores.

Notice also that these four volcanoes are located in the northern hemisphere, the southernmost one being Thera at 36°N longitude while Saint Helens and Aniakchak are above 45°N.

Despite this geographic concentration of volcanic activity in the north, the diagram below shows that the SO_2 spike that appears ca. 1,620 BC is *almost four times larger in the Antarctic ice core – 620 ppm as indicated by the black arrow* – than it is in Greenland ice core –180 ppm as shown by the grey arrow in the diagram on the next page.

So what could explain the prevalence of atmospheric sulfur in the Antarctic while eruptions occurred in the northern hemisphere? What could explain the 58% unaccounted atmospheric sulfur? What could explain the occurrence of, at least, four virtually simultaneous major eruptions?

A cometary event is a prime candidate because direct impacts, along with overhead explosions, can generate a lot of atmospheric dust and, as we will see later, there is a proven correlation between cometary activity and volcanic activity.[445]

Cometary activity is obviously a plausible cause, but is there any trace of such activity in the records of the time? Actually, Chinese writers report such an event:

> According to the Bamboo Annals,[446]"In [king Jie's] 10th year, the five planets went out of their courses. **In the night, stars fell like rain.** The earth shook." Later in the text, the Bamboo Annals state, *The sky was overspread with mists for three days.* The mists occurred during the reign of king T'ae-Këah, the fourth ruler of the Shang dynasty, who was enthroned c. 1530 BC. The reign of king Jie of Xia, which ended in approximately 1600 BC, and the reign of king T'ae-Këah of Shang, which occurred nearly seventy years later **both occurred within the c. 1670–1520 BC radiocarbon date range of Thera and the ice-core and tree-ring range of c. 1740–1440 BC.**[447]

---

[444]Ibid.
[445]See chapter "Correlation between Cometary Activity And Volcanic Activity."

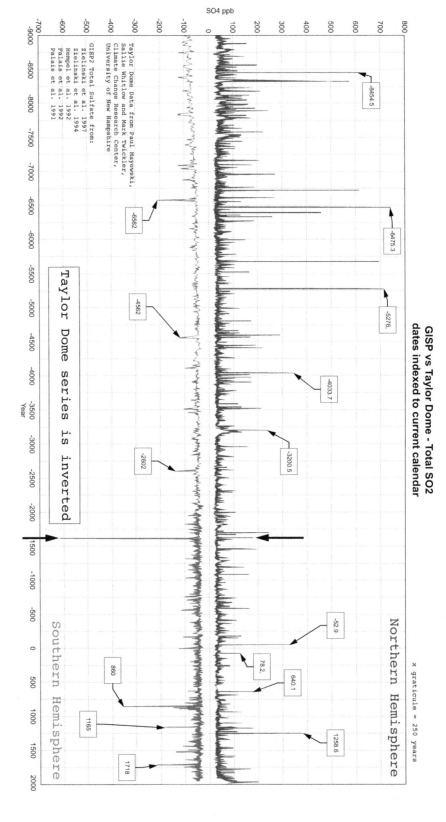

Figure 110: © Volcano Café
SO$_2$ concentration – Greenland vs. Antarctica.

So, the Chinese annals appear to have recorded something akin to a close cometary passage at the time of the 1,600 BC events. Notice also the mention of earthquakes ("the Earth shook"). As explained later,[448] along with volcanic activity, seismic activity is also closely correlated with cometary events.

The "stars falling like rain" as reported by the Bamboo Annals was obviously not a minor event, because 1,600 BC also marked the end of the Xia dynasty,[449] which had ruled over Eastern China for four centuries.

At this point, there is a lot of data to support the occurrence of a cometary-induced catastrophe ca. 1628 BC, the main counter-argument being the Egyptian

Figure 111: © Wiki Commons
Small sample of The Bamboo Annals

records. Indeed, there is not a single trace of a catastrophic event ca. 1628 BC in those documents, although there are catastrophes ascribed to different dates.

But little credence should be given to Egyptians records. Their highly controversial and still debated chronology was designed to match the chronology of the Old Testament, which is, in itself, a bundle of mythical accounts, not real history:

> It's long been known that **most of the OT [Old Testament] is fiction** (Exodus, Job, Ruth) **or forgery** (Daniel, Deutero-Isaiah, Deutero-Zecharia).[450]

Professor Thomas L. Thompson, one of the world's foremost authorities on biblical archaeology, states:

> The **first 10 books of the Old Testament are almost certainly fiction**, written between 500 and 1,500 years after the events they purport to describe.[451]

The first ten books of the Old Testament cover the time span[452] from Genesis to ca. 1,100 BC, therefore including the 17th century BC.

---

[446]The Bamboo Annals, also known as the Ji Tomb Annals or Zhushu Jinian, are Chinese court records written on bamboo strips covering almost five millennia from ca. 5,000 BP to 300 AD. See: Wikipedia Editors (2006), "Bamboo Annals," *Wikipedia*.
[447]James Legge (1865), *The Chinese Classics Vol. III, Part I*, Trübner & Co.
[448]Chapter "Correlation between Cometary Activity And Volcanic Activity."
[449]Pure Insight Editors (2003), "Reflections on History: Natural Disasters and the Decline and Fall of the Xia Dynasty," *PureInsight.com*
[450]Richard Carrier (2014), *On the Historicity of Jesus: Why We Might Have Reason for Doubt*, Sheffield Phoenix Press, p. 215.
[451]Thomas L. Thompson (2000), *Early History of the Israelite People: From the Written & Archaeological Sources*, Brill.
[452]Edwin R. Thiele (1983), *The Mysterious Numbers of the Hebrew Kings*, Kregel Academic.

The 7,200 BP (5,200 BC) Event

Now that we have seen data suggesting the possible occurrence of comet-induced catastrophes ca. 14,400 BP (3,600 x 4) and ca. 3,600 BP, let's further our investigation of an hypothetical 3,600-year comet cycle by looking at the following date: 7,200 BP (3,600 x 2).

Similarly to 14,400 and 3,600 BP, we discover a sharp temperature drop revealed by ice core analysis, as depicted by the black vertical line in the graph below:

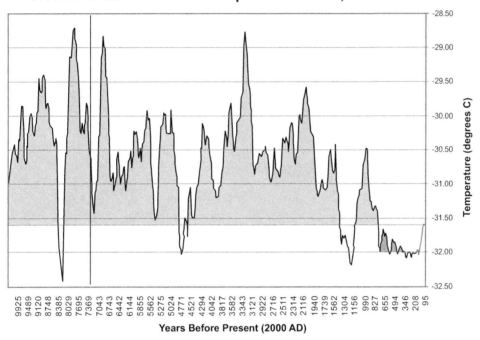

Figure 112: Eastbrook GISP2 ice core temperature reconstruction

7,200 BP is also marked by an exceptional spike in mercury (Hg) concentration. The graph below shows the concentration in titanium, bromine and mercury found in EGR2A, a six-meter-long peat core extracted in Switzerland that spans from 12,500 BP to now.[453] Over those 14 millennia, the largest concentration in mercury occurred *at 7,265 ± 75 BP*, as indicated by the black arrow:

---

[453] F. Roos-Barraclough *et al.* (2002), "A 14 500 year record of the accumulation of atmospheric mercury in peat: volcanic signals, anthropogenic influences and a correlation to bromine accumulation," *Earth and Planetary Science Letters* 202:435–451.

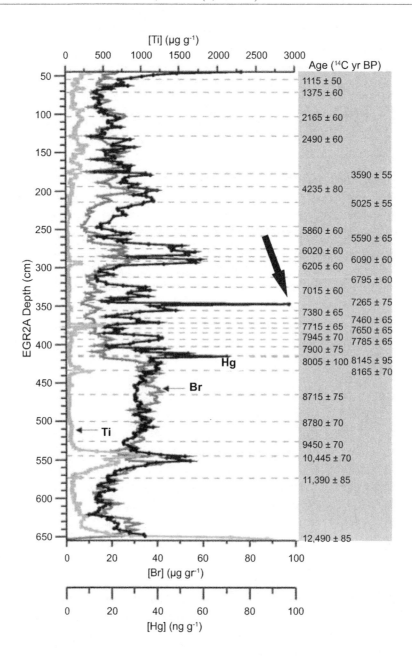

Figure 113: © Roos-Barraclough
Titanium, bromine and mercury concentration in the EGRA2A peat core

Mercury is a very rare element on planet Earth. Its bulk abundance is estimated[454] at 3–8 ppb.[455] However, in meteorites, mercury concentration is much higher, albeit quite variable, with concentrations ranging from around 10 ppb to 14,000 ppb.[456] The record 14,000 ppb concentra-

---

[454]W.F. McDonough et al. (1995), "The composition of the Earth," *Chemical Geology* 120(3–4):223–253.
[455]Parts per billion.
[456]M.M. Meier et al. (2016), "Mercury (Hg) in meteorites: Variations in abundance, thermal release profile, mass-dependent and mass-independent isotopic fractionation," *Geochimica et Cosmochimica Acta* 182:55–72.

tion was found in the Orgueil meteorites.⁴⁵⁷ According to chemistry of the cosmos professor Matthieu Gounelle, the Orgueil meteorites have a cometary origin,⁴⁵⁸ which is also the case for some other carbonaceous chondrite meteorites.⁴⁵⁹

The connection between comets and high mercury concentration is reinforced by the analysis of impact craters on the Moon:

> High Hg [mercury] abundances in these deposits were suggested by Reed (1999), and confirmed in the **plume ejected from the floor of the lunar polar crater** Cabeus by a kinetic impactor (the LCROSS mission; Gladstone et al., 2010). **The lunar soil in this crater was found to contain 2,000,000 ppb Hg!**⁴⁶⁰

The floor of the lunar crater Cabeus exhibits a mercury concentration of 2,000,000 ppb, which is about *300,000 larger* than the average mercury concentration found on Earth.

In addition to an unusual mercury spike and its probable cometary origin, 7,200 BP is also marked by a sudden increase in atmospheric sulfur⁴⁶¹ concentration discovered in Greenland ice cores.

Figure 114: © Eunostos
One of the Orgueil meteorites

In the diagram on the next page, the two black arrows show the 7,200 BP date and massive SO₂ spikes that are the *fifth largest sulfur signals over the past 70,000 years.*

The next diagram on the page after displays the number of major eruptions per century found in two different ice cores: EPICA (Antarctica) and GISP2 (Greenland),⁴⁶² inferred from the sulfate release. As you can see, the 53ʳᵈ century BC was a period of increased volcanic activity.

---

[457] About 60 carbonaceous chondrite meteorites fell around Orgueil, France, in 1864. The total weight was about 14 kg (30 pounds). See: John G. Burke (1986), *Cosmic Debris: Meteorites in History*, University of California Press, pp. 168–169.

[458] Matthieu Gounelle *et al.* (2006), "The orbit and atmospheric trajectory of the Orgueil meteorite from historical records," *Meteoritics & Planetary Science* 41:135–150.

[459] H. Haack *et al.* (2011), "CM Chondrites from Comets? — New Constraints from the Orbit of the Maribo CM Chondrite Fall," *LPI Contributions*.

[460] Meier (2016).

[461] In sulfur dioxide form – SO₂.

[462] A.V. Kurbatov *et al.* (2006), "A 12,000 year record of explosive volcanism in the Siple Dome Ice Core, West Antarctica," *Journal of Geophysical Research* 111(D12).

The 7,200 BP (5,200 BC) Event

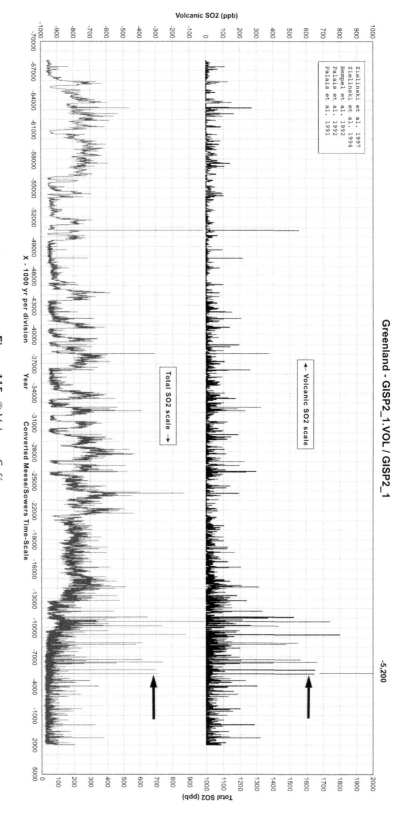

Figure 115: © Volcano Café
SO₂ concentration over the past 70,000 years

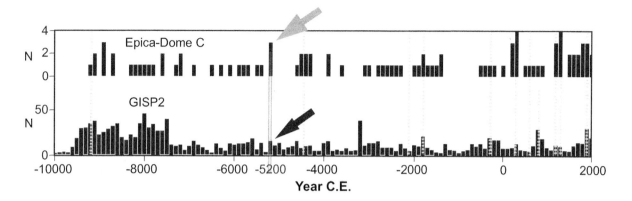

Figure 116: © Sott.net, adapted from Kurbatov
Supposed number of major eruptions per century.

5,200 BC (7,200 BP) reveals a spike in number of eruptions recorded in EPICA that amounts to 3 major events within one century, as shown by the vertical light grey line and the grey arrow in the graph above. The GISP2 ice cores also reveal an activity spike – although not as pronounced – with about 15 major eruptions, as indicated by the black arrow.

If we look in more detail at this period of time, the timing of the sulfate release is as follows:

5,300 to 5,201 BC Century	
Year (BC)	SO4 concentration (ppm)
5209	67
5237	44
5240	34
5245	38
5277	677
5279	404
Century total	**1264**

Figure 117: © Sott.net
SO$_4$ concentration (5279–5209 BP)

The original paper[463] from which this data comes lists only the releases exceeding 25 ppm of sulfate. Most centuries are very quiet, with a total SO$_4$ release below 100 ppm, but the 53rd century BC exhibits more than 1264 ppm.

For comparison, that is about eight times the previously described SO$_4$ signature left by Thera in 1,628 BC or 40 times the 1883 Krakatoa signature.

Russian volcanologist Zielinsky has proposed the Kizimin volcano located in Kamchatka, Russia, as the cause of the 5,277 and 5,279 BP spikes.[464] Nevertheless, the Kizimin hypothesis is

---
[463]Zielinski (1994).
[464]Zielinski (1994), p. 14.

debatable because no tephra analysis[465] has confirmed it yet[466] and the carbon dating of the two consecutive Kizimin eruptions are estimated between 5,600 and 5,000 BC.

The above raises several questions: was Kizimin the sole contributor, or even a contributor at all, to the 7,200 BP sulfate spike? What triggered the wave of virtually simultaneous sulfate releases that marks the 5279–5209 BC interval?

Despite a scarcity of archeological sites for this period of time, one can find the marks left by the 7,200 BP event on human activity. One of those sites is Çatalhöyük, Turkey, which was founded ca. 9,500 BP[467] and flourished for 23 centuries until its abandonment ca. 7,200 BP.[468]

Figure 118: © NiglayNik
Covered excavation site at Çatalhöyük

While Çatalhöyük is the most notable, several other Neolithic sites were abandoned or destroyed around the same time.

In Hacilar, Turkey, a Neolithic fortified village was built on top of a mound. The location was in close proximity to an active geological fault[469] and was destroyed ca. 7,200.[470]

The Halaf culture started ca. 8,000 BP and developed for eight centuries on a vast territory comprising southeastern Turkey, Syria and northern Iraq. The Halaf culture ended ca. 7,200 BP[471] when many Halafians settlements were abandoned.[472] This is also the case, for example, of the Yarim Tepe site in Iraq, which collapsed ca. 7,200 BP after about one millennium of occupation.[473]

---

[465] Tephra analysis allows matching the specific blend of minerals released by a given volcano with the same blend of minerals in a given ice core. Because of the unique mineral blend of each volcano, it provides highly likely matches, unlike analysis solely based on sulfur release, which can be due to any eruption among many other factors, including cometary activity.

[466] Kurbatov (2006), p. 14.

[467] Craig Cessford (2015), "A new dating sequence for Çatalhöyük," *Antiquity* 75:717–725.

[468] Mary Settegast (1987), *Plato Prehistorian*, The Rotenberg Press, p. 207.

[469] Maxime Brami (2014), "Revisiting Hacılar," *Arkeoloji ve Sanat Dergisi* 146(13).

[470] William Irwin Thompson (1996), *The Time Falling Bodies Take To Light: Mythology, Sexuality and the Origins of Culture*, Palgrave Macmillan.

[471] Michela Spataro *et al.* (2010), "Centralisation or Regional Identity in the Halaf Period? Examining Interactions within Fine Painted Ware Production," *Paléorient* 36(2):91–116.

[472] Georges Roux (1992), *Ancient Iraq*, Penguin UK, p. 101.

[473] Stuart Campbell (2007), "Rethinking Halaf Chronologies," *Paléorient* 33(1):103–136.

Figure 119: © StevanB
Excavation of Yarim Tepe

The collapse of Neolithic sites ca. 7,200 BP was not limited to the Middle East, however. The site of Ban Rai, Vietnam, is a large rock shelter that was used by humans from 10,600 BP until its abandonment in 7,200 BP after 34 centuries of activity.[474]

The Chertovy Vorota Cave[475] in Russia – occupied by humans for 22 centuries – was also abandoned at this time. Similarly, in China, the culture of Xinglongwa is the first sedentary culture in Asia.[476] It flourished for ten centuries before ending ca. 7,200 BP.[477]

---

[474] Vicki Cummings *et al.* (2014), *The Oxford Handbook of the Archaeology and Anthropology of Hunter-Gatherers*, OUP Oxford.

[475] Wikipedia Editors (2017), "Chertovy Vorota Cave," *Wikipedia*.

[476] Gideon Shelach (2000), "The Earliest Neolithic Cultures of Northeast China: Recent Discoveries and New Perspectives on the Beginning of Agriculture," *Journal of World Prehistory* 14:363–413.

[477] Xin Jia *et al.* (2017), "Spatial and temporal variations in prehistoric human settlement and their influencing factors on the south bank of the Xar Moron River, Northeastern China," *Frontiers of Earth Science* 11(1):137–147.

The 10,800 BP (8,800 BC) Event

At this point we have examined the 14,400 BP (4 x 3,600), 3,600 BP and 7,200 BP (2 x 3,600) events. Only one date remains: the 10,800 BP event (3 x 3,600).

Just like with the three preceding dates, we notice a sudden drop in temperatures around 10,800 BP:

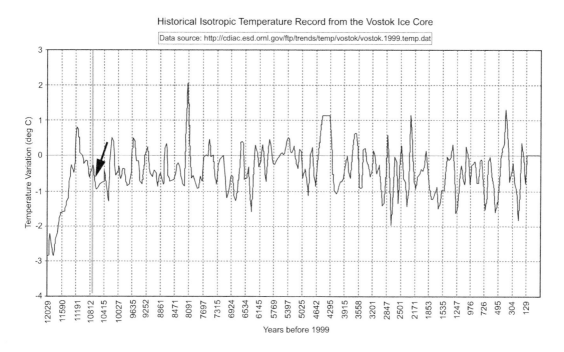

Figure 120: © CDIAC
Vostok temperature reconstruction

The diagram above is a temperature reconstruction based on the analysis of Vostok, Antarctica, ice cores. We can see a black arrow and a grey vertical line showing a temperature drop ca. 10,800 BP. Notice that this temperature drop is moderate – about 0.7°C – compared to the three other events. Indeed, most of the drops in the record are much more pronounced, so we must look for other signals in the data to confirm if this was indeed an anomaly.

Reconstructed temperature from a stalagmite in the Dongge cave, China, reveals the same 0.7°C temperature drop ca. 10,800 BP,[478] as indicated by the black and grey vertical line in the illustration below:

---

[478] Carolyn Dykoski (2005), "A high-resolution, absolute-dated Holocene and deglacial Asian monsoon record from Dongge Cave, China," *Earth and Planetary Science Letters* 233(1–2):71–86.

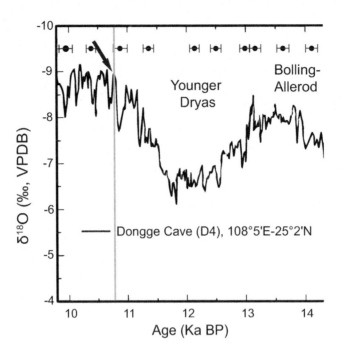

Figure 121: © Dykoski
Temperature reconstruction from Dongge Cave

However, this drop in temperature does not appear in the Greenland ice core, as shown by the black vertical line in the graph below:

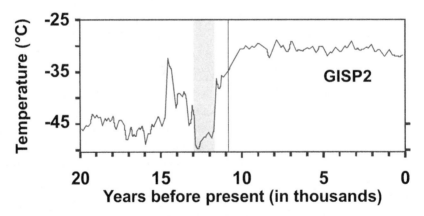

Figure 122: © Alley
Temperature in Greenland 20 kY BP to now

The temperature drop in Antarctica and China combined with the absence of a temperature drop in Greenland suggest that if the 10,800 BP event was caused by a cometary impact, it happened *in the southern hemisphere*.

Temperature is not the only parameter that markedly fluctuated ca. 10,800 BP, which also reveals a sudden spike in nitrate (NO_3) concentration, as indicated by the black arrows and black vertical lines below:

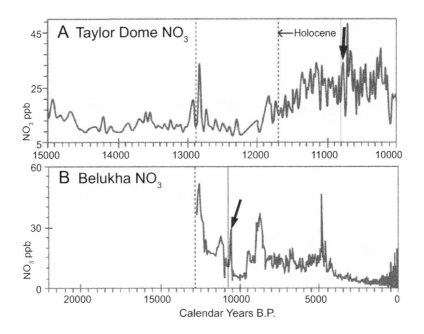

Figure 123: © Wolbach
NO3 concentration at Taylor Dome and Belukha

As shown in the diagram above, both Taylor Dome, Antarctica, and Belukha, Siberia, experienced a spike in nitrate ca. 10,800 BP. Nitrate is known to be a biomass burning proxy.[479]

Other typical combustion aerosols are acetate oxalate, NH_4[480] and formate.[481] Coincidentally or not, *each of the these four chemicals displays a marked rise starting from 10,800 BP*, illustrated in the graph to the right by the black vertical lines, while the black arrows indicate the subsequent concentration spikes.

This increase is not as pronounced at the onset of the Younger Dryas ca. 12,900 BP, but it is still identifiable.

The diagram below shows the concentration of titanium (Ti – light grey curve), bromine (Br – medium gray curve) and mercury (Hg – black curve) found in the previously mentioned EGR2A peat

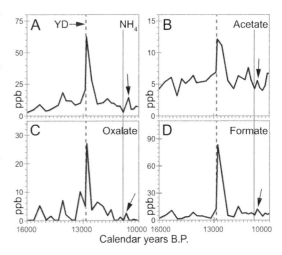

Figure 124: © Wolbach
Combustion aerosols in the GRISP ice core

---

[479] Ed Brook *et al.* (2015), "Isotopic constraints on greenhouse gas variability during the last deglaciation from blue ice Archives," in Michael Sarnthein *et al.*, *Deglacial changes in ocean dynamics and atmospheric CO2: modern, glacial, and deglacial carbon transfer between ocean, atmosphere and land*, Leopoldina Symposium, Nova Acta Leopoldina, pp. 39–43.

[480] NH_4 is the formula for the ammonium cation, a positively charged ion whose full chemical formula is NH 4.

[481] Wendy Wolbach *et al.* (2018), "Extraordinary Biomass-Burning Episode and Impact Winter Triggered by the Younger Dryas Cosmic Impact ∼12,800 Years Ago," *The Journal of Geology* 126(2).

core.[482] It reveals one of the largest concentrations of titanium (light grey arrow), bromine (medium grey arrow) and mercury (black arrow) ca. 10,800 BP:

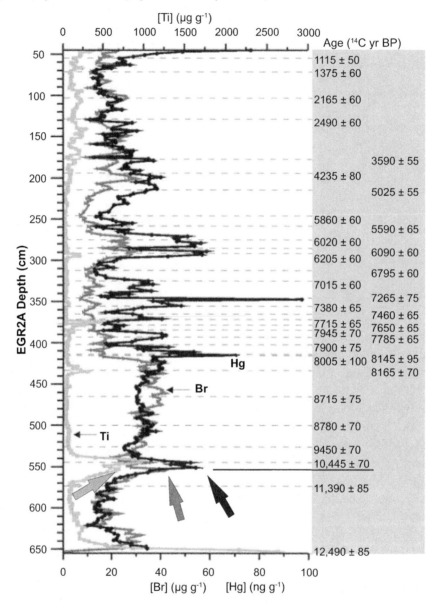

Figure 125: © Roos-Barraclough
Titanium, Bromine and mercury concentration in peat core

In the previous chapter, we described the high concentration of mercury found in meteorites and impact craters. Asteroids are also rich in titanium.[483] In fact, as shown in the table below, titanium is one of the very few elements found in every single class of meteorite.[484] Its concentra-

---

[482]See previous chapter, "The 7200 BP (5200 BC) Event."
[483]Aparna Kher (2020), "10 Things You Need to Know about Asteroids," *TimeAndDate.com*
[484]Brian Mason *et al.* (1970), "Minor and Trace Elements in Meteoritic Minerals," *Smithsonian Contributions to*

tion is particularly high in the following classes: Orthopyroxene, Clinopyroxene, Chromite and Phosphate, as indicated by the four grey boxes below:

Table 4. – Titanium (ppm)

Meteorite	Metal	Troilite	Olivine	Ortho-pyroxene	Clino-pyroxene	Plagioclase	Chromite	Phosphate	Bulk
Camperdown			140	350	1000				
Modoc	< 50	50	100	1000		360	17000	2100	700
St. Severin	< 50	50	220	1000		650	19000	1200	650
Winona				1200	3800	200			840
Haraiya					2100	340			
Marjalahti			46						
Springwater			40						
Johnstown				1200					
Mt. Egerton				120					
Soroti			22						

Figure 126: © Mason
Titanium concentration according to meteorite classes

According to EGR2A peat core analysis, 10,800 BP also marks the largest concentration spike in bromide[485] over the past 12,500 years. Bromide concentration in Earth's crust is low, about 1 ppm,[486] which is similar to the bromide concentration found in meteorites.[487]

If there's not much bromide in meteorites and in Earth's crust, where could the 40 ppm bromide spike found ca. 10,800 BP come from?

With 65 ppm,[488] the *ocean* exhibits much higher bromide concentration than the Earth's crust. However, the 40-ppm bromide spike was not found on sea shores but in a peat core collected at Etang de la Gruere, Switzerland, hundreds of kilometers from the nearest coast.

Could an asteroid impact in the ocean explain this unexpected bromide spike found in Switzerland? The research conducted by impact modeling expert Elisabetta Pierazzo suggests that asteroid impacts in the ocean increase the atmospheric bromide concentration 20 times and can spread this atmospheric bromide *over whole continents*:

> The impact of a 500 m asteroid [in the ocean] increases the upper atmospheric water vapor content by more than 1.5 times the background over a wide region surrounding the impact point for the first month after the impact. **Halogens, ClY**[489]**and BrY,**[490]**follow the water vapor distribution, with an initial increase of over 20 and 5 times normal background**, respectively, in the same region surrounding the impact. The perturbations eventually **spread over the northern hemisphere, where water vapor content remains about 50% above background for the first year after impact, while ClY and BrY exceed five times and twice their background values.**[491]

---

the Earth Sciences: 1–17.

[485] Bromide is the reduced form, i.e. having gained one electron, of the element bromine (Br).

[486] Wikipedia Editors (2007), "Abundance of elements in Earth's crust," *Wikipedia*. The 1 ppm is deduced from the three estimates provided in the table. 0.27 ppm for *Barbalace*, 3 ppm for *WebElements* and 2.4 ppm for *CRC Handbook of Chemistry and Physic*.

[487] M. Langenauer et al. (1993), "Depth-profiles and surface enrichment of the halogens in four Antarctic H5 chondrites and in two non-Antarctic chondrites," *Meteoritics* 28(1):98–104.

[488] H.U. Sverdrup et al. (1942), *The Oceans, Their Physics, Chemistry, and General Biology*, Prentice-Hall, chapter VI: "Chemistry of Sea Water."

One or more oceanic impacts would explain the large bromide spike *and* the moderate and heterogeneous temperature drop, because oceanic impacts generate far less atmospheric dust than ground impacts.

The relatively low concentration of atmospheric dust c. 10,800 BP is confirmed by the diagram below. The 10,800 BP signal is shown by the grey arrow, with a SO_4 level reaching about 150 ppm. This is not negligible compared to the average background concentration of 25 ppm, but *is* modest relative, for example, to the previously described 7,200 BP event, which was marked by a 750 ppm SO_4 spike, as indicated by the black arrow below:

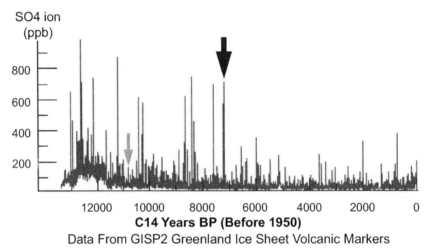

Figure 127: SO_4 concentration in GISP2 ice core

Nonetheless, some of the 10,800 BP impacts might have also hit the crust, as suggested by the previously described spikes in NO_3, NH_4, oxalate, formate and acetate, which are proxies for biomass burning. The event that left its mark on Earth ca. 10,800 BP triggered massive biomass burning and also directly impacted fauna:

> **An extensive but discontinuous charcoal layer, dated to 10,840 ± 80 and 10,960 ± 60 BP**, may mark the arrival of fire-starting humans. **Whether or not the burning was anthropogenic**, above this lens, two beveled-based bone points, a reworked Gainey like nuted point, and two possibly human-altered bones (a cut snapping turtle vertebra and a perforated ilium of a peccary) were found within a ca. 2-m² area. Based upon overlying dates (unfortunately including a few anomalies due to evident water disturbance), the cultural material seems to date to ca. 10.800– 10,900 BP. The dated samples from the overlying strata include two bones of the extinct flat-headed peccary (*Piatygouus compressus*) (11,130 ± 60 and 11,060 ± 60 BP) a bone of extinct giant beaver (*Castoroides ohioeusis*) (10.850 ± 60 BP), and a bone of caribou (*Rangifer taraudus*) dated to 10.440 ± 40 BP. The latter is obviously not an extinct species but it's been quite a while since caribou lived in Ohio.
>
> **Several caves in the Southwest** contain stratified dated sequences that are merely **suggestive of a Terminal Pleistocene catastrophe**, because no evidence of human habitation or

---

[489] Chlorine compounds.
[490] Bromine compounds.
[491] Elizabetta Pierazzo *et al.* (2010), "Ozone perturbation from medium-size asteroid impacts in the ocean," *Earth and Planetary Science Letters* 299(3–4):263–272.

butchery is present. In these sites, **steady deposition of dung through millennia by Shasta ground sloth (*Nothrotheriops shastensis*), stops abruptly at 11,000–10,800 BP.** Relatively precise terminal dates on dung include: Gypsum Cave, NV, 11,005 ± 100, 11,080 ± 90 BP; Rampart Cave, AZ, 10.940 ± 60, 11,000± 140 BP; Muav Caves, AZ, 11140± 160, 11.060±240, 10,650 ± 220 BP; Aden Crater, NM, 11.080 ± 200 BP; Upper Sloth Caves, TX, 10,750 ± 140, 10,780 ± 140 and 11.060 ± 180 BP ...[492]

As emphasized by geologist Robert Thomson in his concluding words about the extinction of the above-mentioned Shasta ground sloth, there was no evidence of dietary stress at the time of extinction,[493] which *suggests a sudden demise of a catastrophic nature.*

Figure 128: © David Craig
Artist rendition of the Shasta ground sloth.

---

[492] Gary Haynes (2008), *American Megafaunal Extinctions at the End of the Pleistocene*, Springer, p. 30.
[493] Robert Thompson *et al.* (1980), "Shasta ground sloth (*Nothrotheriops shastensis hoffstetter*) at Shelter Cave, New Mexico: Environment, diet, and extinction," *Quaternary Research* 14(3):360–376.

Year C.E. (Duration, years)	Signal, ppb	Volcanic Eruptions Possibly Associated With Detected Events	Year C.E. (Duration, years)	Signal, ppb	Volcanic Eruptions Possibly Associated With Detected Events
1975 (2.8)	44 {43}	Fuego, Guatemala 1974 (VEI = 4)	1100 (2.7)	45 {40}	no match found
1964 (1.9)	{42}	Agung, Indonesia 1963 (VEI = 5)	1061 (2.7)	{37}	no match found
1936 (1.2)	83	Bristol Island, Antarctica 1936 (VEI = 2)	1052 (2.3)	45 {49}	no match found
1920 (3.1)	74 **(79)**	Puyehue, Chile 1921 (VEI = 4)	1035 (2.5)	55 {43}	Cerro Hudson, Chile, cal 2σ C.E. 655(1004)1159
1889 (2.2)	62 **(66)**	Lonquimay, Chile 1887–1889 (VEI = 3)	1016 (2.4)	84 **{57}**	no match found
1887 (2.1)	{42}	Tarawera, New Zealand 1886 (VEI = 5)	991 (2.6)	41 {46}	no match found
1883 (1.7)	{37}	Krakatau, Indonesia 1883 (VEI = 6)	958 (2.9)	44 {54}	no match found
1876 (3)	{41}	Cotopaxi, Ecuador 1877 (VEI = 4)	879 (3.1)	50 {52}	Cotopaxi, Ecuador, soil below, cal 2σ C.E. 665(885)1019
1871 (2.6)	{38}	Deception Island, Antarctica 1871	857 (2.5)	{38}	no match found
1839 (1.9)	55 {58}	Buckle Island, Antarctica 1839	847 (2.8)	48 {46}	Cerro Bravo, Columbia, 2σ cal C.E. 562(790,842,859)1160
1831 (1.9)	59 {47}	Hodson, Antarctica 1831	837 (2.6)	95 **{85}**	no match found
1809 (3.1)	99 **{72}** A	Buckle Island, Pleiades, Antarctica	828 (2.8)	64 {51}	no match found
1805 (3.0)	T	no match found	820 (2.2)	63	no match found
1804 (1.0)	A	Buckle Island, Antarctica	773 (3)	42 {40}	El Chichon, Mexico, 1σ cal C.E. 675(779,788)961
1787 (3.2)	84 **{68}**	no match found	723 (2.6)	{37}	no match found
1746 (2.6)	{37}	Cotopaxi, Ecuador 1744 (VEI = 4)	702 (3)	51 {44}	no match found
1724 (3.2)	91 {56}	no match found	695 (2.3)	44 {38}	no match found
1707 (3.1)	59 {38}	Cerro Hudson, Chile, cal 2σ C.E. 1447(1732)1953	640 (2.6)	43	Rabaul, Papua New Guinea, (VEI = 6), 2σ cal C.E. 427 (646) 804
1614 (1.9)	{39}	no match found	607 (2.1)	{40}	El Chichon, Mexico, 1σ cal C.E. 441(553,600,614)662
1601 (1.9)	61	Huaynaputina, Peru 1600 (VEI = 6)	605 (1.9)	43	no match found
1592 (2.1)	**91 {60}**	Raung, Java 1593 (VEI = 5)	590 (1.9)	86 **{87}**	El Chichon, Mexico, 1σ cal C.E. 256(385, 439, 459, 478, 510, 531)654
1569 (2)	62 {46}	Billy Mitchell, Papua New Guinea, 2σ cal C.E. 1451(1539)1627 (VEI = 6)	497 (2.7)	{39}	no match found
1471 (1.9)	50	no match found	443 (2.3)	67 **{68}**	no match found
1462 (2.8)	47	El Misti, Peru 1438–1471(VEI = 3+)	423 (2.6)	{37}	Ilopango, El Salvador, 1σ cal C.E. 421(429)526
1459 (1.8)	60 {45}	no match found	409 (2.3)	{43}	Krakatau, Indonesia 416 C.E.
1450 (2.1)	**97 {80}**	no match found	404 (2.2)	{43}	Cerro Hudson, Chile, 390 C.E.
1448 (2.1)	55	Kuwae, Vanuatu, 145 ± 10 (VEI = 6)	344 (2.1)	51 {43}	Melimoyu, Chile 350 ± 200 C.E.
1445 (3.1)	70 {37}	no match found	335 (3)	71	no match found
1443 (2.5)	84	no match found	306 (2.8)	90 **{60}**	no match found
1412 (2.1)	55	Pinatubo, 1450 ± 50 (VEI = 5)	304 (2.3)	**125 {98}** A	Mount Melbourne
1376 (2.6)	90 **{65}**	no match found	290 (2.2)	80	no match found
1368 (2.5)	49	El Chichon, Mexico, 1σ cal C.E. 1320(1408)1433	279 (2.3)	45 {41}	Cotopaxi, Ecuador, soil below, cal 2σ C.E. 4(245,310,315)537
1346 (2.2)	53 {38}	Cerro Bravo, Columbia, 2σ cal C.E. 264(1305, 1365, 1386)1436	275 (1.7)	78 **{74}**	no match found
1278 (2.7)	108 **{90}**	no match found	267 (1.6)	61 {40}	no match found
1271 (2.7)	139 **{108}**	no match found	265 (2.3)	70 **{68}**	no match found
1262 (3)	163 **{133}**	Mount Melbournek	262 (2.6)	54 {47}	no match found
1259 (2.6)	45 {43}	Quilotoa, Ecuador, 2σ cal C.E. 1160(1260)1360	259 (2.7)	52 {44}	no match found
1234 (2)	58 {44}	Cotopaxi, Ecuador, 2σ cal C.E. 1024(1221)1376	252 (2.6)	113 **{86}**	no match found
1193 (2.5)	117 **{91}**	no match found	239 (2.1)	{41}	El Chichon, Mexico, charcoal, 1σ cal C.E. 132(249)394
1186 (2.2)	43	El Chichon, Mexico, 1σ cal C.E. 1025(12201)1278	190 (2)	{37}	Pico De Orizaba, Mexico, 2σ cal C.E. 67(131)243
1183 (2.5)	{37}	Cerro Bravo, Columbia, 2σ cal C.E. 996(1160)1279	176 (2.2)	59 **{61}**	Taupo, New Zealand 180 C.E. (VEI = 6+)
1175 (2.7)	78 **{74}**	no match found	94 (2.3)	43	Cotopaxi, Ecuador, soil below, 2σ cal B.C.E. 350(cal C.E. 128)cal C.E. 534
1172 (2.9)	114 **{86}**	no match found	91 (2.2)	51 {53}	no match found
1132 (2.7)	52 {48}	no match found	67 (2.8)	43 {41}	no match found
1111 (2.7)	{42}	no match found	21 (2)	{37}	no match found
1100 (2.7)	45 {40}	no match found	47 (2)	{38}	Tacana, Mexico/Guatemala, 1σ cal B.C.E. 38 (cal C.E. 25–72)cal C.E. 216

Figure 129: © Kurbatov
List of SO₄ signals and possibly associated volcanic eruptions

Mystery Eruptions

Sulfur spikes have been systematically found in our analysis of the 14,400, 10,800, 7,200 and 3,600 BP events. From the point of view of mainstream science, these spikes are *exclusively a proxy for volcanic eruptions*. That is, if a sulfur spike is found, it can only be caused by a specific volcanic eruption.

The problem, however, is that even in the recent past, most sulfur spikes remain unexplained because there are *no matching volcanic eruptions* in the records. The diagram below lists the 92 notable sulfur spikes that occurred over the past 2,000 years. As indicated by the grey horizontal lines in the table on the previous page, **more than half of them**, 47 spikes out of 92, are not attributed to any eruption and the remaining 45 spikes are only 'possibly associated' with a given eruption[494].

The further back we go in time, the worse it gets. The diagram below shows SO_2 concentrations in the GISP2 ice core over the past 16,000 years. It shows 62 spikes – as indicated by the black arrows – of more than 120 ppm. For comparison the 'giant' 1883 Krakatoa eruption generated about 40 ppm of sulfur.[495]

Out of the 62 major spikes, only 15 are *tentatively* associated with volcanic eruptions. So, *more than 75% of the sulfur spikes are not associated with any volcanic eruption whatsoever*, including the two largest ones occurring ca. 10,657 BC and 9,285 BC. Those dates are highlighted in thick black boxes on the page following the table.

Of course some of these mystery spikes are indeed due to volcanic eruptions and it's only a matter of time before the culprit is identified. For a match to be ascertained the timing of the eruption has to obviously match the timing of the sulfur spike, but more importantly, the specific geochemical composition of the eruption tephra has to match the geochemical composition of the ice core section where the sulfur spike occurred.[496]

A case in point is the largest sulfur spike in written history. In the diagram above we can see the grey-highlighted 1258 AD spike which reaches almost 400 ppm, or ten times the emission of the 1883 Krakatoa eruption. For decades this spike was not associated with any volcanic eruptions:

> The **largest eruption of the historic period**, possibly of the past 7000 years, probably occurred in AD 1257. Its estimated magnitude (10^{14}–10^{15}kg) and caldera diameter (10–30 km) make it **surprising that the volcano responsible has not been identified**, but investigations of candidate young calderas, and the marine core record, may ultimately reveal the source.

---

[494]Kurbatov (2006).
[495]See part 3, chapter "Sulfur Dioxide."
[496]Kurbatov (2006), pp. 4–8.

Mystery Eruptions

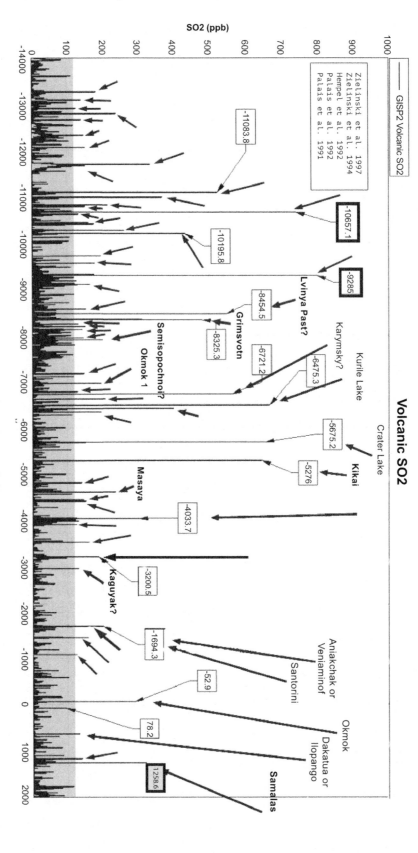

Figure 130: © Volcano Café
SO_2 spikes and possible associated eruptions over the past 16,000 years

Palaeoclimate reconstructions indicate austral and boreal **summer cooling in AD 1257–59, consistent with a high Sulphur yield,** low-latitude explosive eruption occurring in AD 1257. The 5×10^{13}kg Baitoushan eruption probably took place ca. AD 1030. **Other large, unidentified, climate-forcing eruptions occurred in ca AD 1100, 1171, 1229 and 1341.**[497]

The event was so powerful that it may have reduced global temperatures by approximately 2 °C.[498] Some scientists even believe that this event was one of the causes of the Little Ice Age.[499]

Over the years, a few volcanoes had been suspected (Chichon, Quilotoa, Harrat Rahat),[500] but the dating of these eruptions[501] and/or their geochemical profile[502] didn't match.

The AD 1258 spike remained a 'mystery eruption' until 2013 when French geoscientist Frank Lavigne found that the Samalas volcano in Indonesia was the likely culprit based on old Javanese written records and the dating of a tree near the volcano lava deposits.[503]

Figure 131: © TheWanderlovers
Samalas volcano today

---

[497] Clive Oppenheimer (2003), "Ice core and palaeoclimatic evidence for the timing and nature of the great mid-13th century volcanic eruption," *International Journal of Climatology* 23:417–426.

[498] 3.6°F. Sebastien Guillet *et al.* (2015), "Toward a more realistic assessment of the climatic impacts of the 1257 eruption," *EGU General Assembly* 17:1268.

[499] Period of cooling in the northern hemisphere spanning from 1300 to 1850. See, for example: Gifford H. Miller *et al.* (2012), "Abrupt onset of the Little Ice Age triggered by volcanism and sustained by sea-ice/ocean feedbacks," *Geophysical Research Letters* 39(2).

[500] Erik Klemetti (2012), "The Mysterious Missing Eruption of 1258 A.D.," *Wired.com*.

[501] Franck Lavigne *et al.* (2013), "Source of the great A.D. 1257 mystery eruption unveiled, Samalas volcano, Rinjani Volcanic Complex, Indonesia," *Proceedings of the National Academy of Sciences of the United States of America* 110(42).

[502] Alexandra Witze (2012), "Earth: Volcanic bromine destroyed ozone: Blasts emitted gas that erodes protective atmospheric layer," *Science News* 182(1):12.

[503] Lavigne (2013).

Lavigne's hypothesis was confirmed in 2016 by similar glass shards found in the relevant ice cores and in the eruption deposits of Samalas,[504] and in 2019 when tephra found in 1,258 AD ice core sections and eruption products of Samalas revealed a strong geochemical similarity.[505]

But, in most cases, a sulfur spike is attributed to a volcanic eruption simply by matching the dating of dust found in the ice core with the dating of the suspected eruption. This approach doesn't prove that the eruption caused the dust spike, just that the dust spike and the eruption were 'relatively' synchronous.

I use the term 'relatively' because the carbon dating of eruptions through the analysis of the successive lava layers is approximate and exhibits about a 5% uncertainty margin.[506] This means that eruption carbon dated back to 10,000 BP actually happened within a 90% certainty between 10,250 BP and 9,750 BP. That's a five-century uncertainty margin and it's only 90% sure that the eruption falls within this time bracket.

So, the 15 'explained' spikes out of the 65 spikes listed over the past 16,000 years are just hypothetical explanations, where an eruption dating was 'close enough' – we're talking centuries of uncertainty – to the dating of a spike. And even if the identified eruption indeed contributed to the spike, *nothing proves it was the main, much less the sole contributor of the spike*.

A case in point is the 1628 BC event. For a long time Thera was considered the sole culprit. As described previously,[507] thanks to the historical interest and time proximity of the event, further research was conducted involving detailed analysis of the dust found in ice cores that showed that at least three other volcanoes were involved.

In the end, Thera was not the sole contributor at all and there were other unknown contributors besides the four eruptions recently identified.

Another example is the onset of the Younger Dryas ca. 12,900 BP. For years most scientists considered it a purely volcanic event. Despite the overwhelming evidence of cometary bombardments, it is still attributed today by some researchers to volcanic activity exclusively.[508] The confusion between major volcanic eruptions and cometary events is understandable because their 'macro' markers are quite similar:

- A spike in atmospheric aerosols (SO_2, SO_4, NH_4, NO_3, ...), where most of the aerosols are proxies for the burning of biomass, whether comet or volcano induced.

- A sudden temperature decrease induced by the above-mentioned increase in atmospheric dust.

The identification of massive cometary bombardment as the main trigger of the Younger Dryas ca. 12,900 BP is due to the discovery of craters combined with their precise dating and their detailed examination that revealed some specific cometary material likes microspherules, fullerene, platinum, titanium, carbon glass, iridium and nanodiamonds.[509]

---

[504] Anthony Reid (2016), "Revisiting Southeast Asian History with Geology: Some Demographic Consequences of a Dangerous Environment," in Greg Bankoff *et al.*, *Natural Hazards and Peoples in the Indian Ocean World*, Palgrave Macmillan US, p. 33.

[505] Biancamaria Narcisi *et al.* (2019), "Multiple sources for tephra from AD 1259 volcanic signal in Antarctic ice cores," *Quaternary Science Reviews* 210:164–174.

[506] Zielinsky (2006).

[507] See chapter "The 3600 BP (1,600 BC) event."

[508] J.U.L. Baldini *et al.* (2018), "Evaluating the link between the sulfur-rich Laacher volcanic eruption and the Younger Dryas climate anomaly," *Climate of the Past* 14(7):969–990.

[509] See Part I, chapter "The Crime Scene."

Since most of these materials are only found in impact craters and not in volcanoes, scientists had to dismiss the volcano-only hypothesis.

But this raises another important question: *how many dust spikes/temperature drops are wrongly attributed solely to volcanic eruptions?*

Worryingly, the mystery eruptions, i.e. a dust spike to which no eruption is attributed, don't apply only to events long ago. One of those 'mystery eruptions' occurred as recently as 1808. The data is clear, a sulfate spike started around December 1808, as shown by the black vertical arrow in the diagram below:

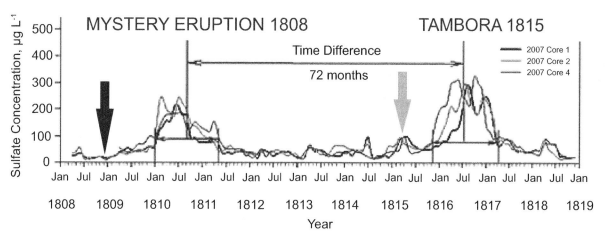

Figure 132: © Cole-Dai
Sulfate concentration in ice core (1808–1819)

Notice that the spike starting around December 1808 is almost as tall as the spike left by the 1815 Tambora eruption, which was the second most powerful eruption in human recorded history, after the previously described Samalas eruption. The 1815 Tambora eruption released more than 300 ppm and was attributed a Volcanic Explosivity Index of 7.

Logically, the marked dust spike observed in December 1808 induced a temperature drop:

> The longest sustained cold period in recent centuries occurred in the early nineteenth century, following the eruption of Tambora and **a second, unidentified but presumably tropical, volcano**. [...Over the past four centuries] tropical anomalies following 1809 are ranked 17th (-0.67°C,[510] 1810) and sixth (-0,77°C,[511] 1811) ...[512]

Beside the above-mentioned sulfur spike and temperature drop, the diagram below reveals a sudden increase in magnesium (grey arrow) and nitrate (black arrow) at the end of 1808.

A spike in nitrate is a proxy for biomass burning, and a similar sudden increase in nitrate was noted during the cometary bombardments that triggered the Younger Dryas ca. 12,900 BP.[513]

As for the magnesium spike, keep it in mind, we'll address it soon.

---

[510] -1.4°F.

[511] -1.5°F.

[512] R. D'Arrigo *et al.* (2009), "The impact of volcanic forcing on tropical temperatures during the past four centuries," *Nature Geoscience* 2(1):51–56.

[513] P.A. Mayewski *et al.* (1993), "The Atmosphere During the Younger Dryas," *Science* 261:195–197.

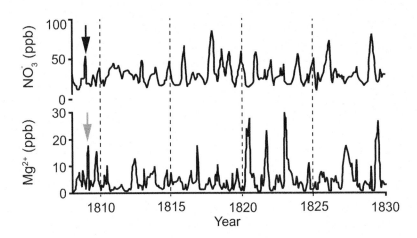

Figure 133: © Steig
Magnesium and nitrate concentrations from ITASE (Antarctica) ice core

Because of its relatively recent occurrence, there are eyewitness reports of the 1808 event. Francisco José de Caldas, director of the astronomical observatory of Santa Fe de Bogotá, Colombia, reported the following:

> As of 11 December of last year [1808], **the disk of the sun has appeared devoid of irradiance**, its light lacking that strength which makes it impossible to observe it easily and without pain. Its natural fiery colour has changed to that of silver, so much so that **many have mistaken it for the moon**. This phenomenon is very noticeable at sunrise, and particularly when the sun sets. When [the sun] is at its zenith, it shines more brightly and cannot be looked at with the naked eye. Near the horizon, it has been seen to take on a light rosy hue, [or] a very pale green, or a blue-grey close to that of steel. ... **The whole vault of the sky has been covered by a light cloud as widespread as it is transparent**. ... [Also] missing have been the emphatic coronas which are so frequently seen around the sun and the moon when those clouds that meteorologists know by the name of veil are present. The stars of the first, second and even the third magnitude have appeared somewhat dimmed, and those of the fourth and fifth have completely disappeared, to the observer's naked eye. **This veil has been constant both by the day and by night**.[514]

The veil event was not limited to Colombia; it was also reported by physician José Hipólito Unanue, who was located in Lima, Peru, 2,600 km[515] from Bogota:

> At sundown in the middle of the month of December [1808], there began to appear towards the S.W., between cerro de los Chorrillos and the sea, an **evening twilight that lit up the atmosphere**. From a N.S. direction on the horizon, it rose towards its zenith in the form of a cone, [and] shone with a clear light until eight [o'clock] at night, when it faded. **This scene was repeated every night until the middle of February**, when it vanished.[516]

Notice that *nowhere is an eruption mentioned*. Despite this, until now, a volcanic eruption remains the sole hypothesis despite the absence of a fitting candidate. Indeed, Urzelina volcano,

---

[514] A. Guevara-Murua *et al.* (2014), "Observations of a stratospheric aerosol veil from a tropical volcanic eruption in December 1808: is this the Unknown ~1809 eruption?" *Climate of the Past* 10(5):1707–1722.
[515] 1,600 miles.
[516] Guevara-Murua (2014).

Tall volcano[517] and Putana volcano, which had been considered as the possible culprits, have eruption dates that don't match the timing of the observed atmospheric veil.[518]

So we can say the 1808 mystery event involved:

- A spike in three atmospheric aerosols: sulfur, nitrate and magnesium.
- A subsequent temperature drop.
- An atmospheric veil spanning at least 2,600 km.

These three features are typical of an overhead cometary explosion or an intense meteor shower. Coincidentally or not, three comets were observed in 1808: C/1808 M1, C/1808 F1 and 26P/1808 C1.[519] They all were discovered by French astronomer Jean Louis Pons from Marseille observatory.

The latter of these three comets is also known as comet Grigg-Skjellerup, whose nucleus is estimated to be 2.6 kilometers in diameter.[520] Grigg-Skjellerup is a periodic comet exhibiting about a five-year cycle, whose perihelion[521] is about 1 AU[522] from the Sun.

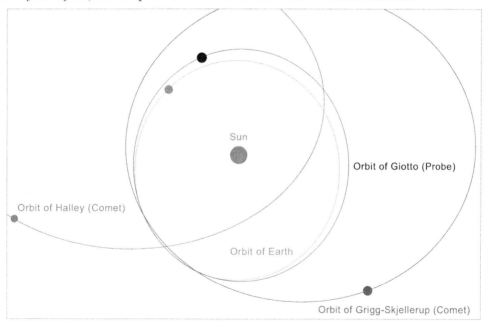

Figure 134: © Wikimedia Commons
Orbits of Earth, Grigg, Giotto and Halley

---

[517] Ulrich Knittel (1999), "History of Taal's Activity to 1911 as Described by Fr. Saderra Maso," Institut für Mineralogie und Lagerstättenlehre, RWTH Aachen University.
[518] Albert Zijlstra (2016), "1809: The missing volcano," *Volcano Café*.
[519] Gary Kronk (1999), *Cometography: Volume 2, 1800–1899: A Catalog of Comets*, Cambridge University Press, p. 9.
[520] 1.6 miles. Ryan S. Park (2007), "JPL Small-Body Database – 26P/Grigg-Skjellerup," Jet Propulsion Laboratory, NASA.
[521] Point in the orbit that is nearest to the Sun.
[522] "AU" stands for Astronomical Unit. It is equal to the distance between the Sun and the Earth.

Having its perihelion so close to Earth's orbit made it an easy target for the *Giotto* space-probe mission in 1992,[523] whose closest approach to Grigg-Skjellerup was only 200 km,[524] as shown in the diagram above, where the orbit of Probe *Giotto* is in black and the orbit of Comet Grigg is in dark grey.

Comet Grigg produces a meteor shower called Pi Puppids[525] that hits our atmosphere every five years during Grigg's perihelion.

Figure 135: © International Meteor Organization
Pi Puppids meteor shower

The picture above shows the latest and rather tame occurrence of the Pi Puppids in 2018. But in the past it seems that the Pi Puppids was a more intense meteor shower.[526]

Now, remember the above-mentioned magnesium spike found in Antarctic ice cores from December 1808 AD?

Interestingly, magnesium is the most abundant element found in some comets. During its flyby near Comet Halley, space probe *Giotto* measured the composition of the comet through spectroscopy. The results revealed that magnesium was by far the most abundant element,[527] as shown by the diagram below where the magnesium spikes are circled in black:

---

[523] ESA Editors (2011), "Giotto, ESA's first deep-space mission: 25 years ago," European Space Agency.
[524] 124 miles.
[525] MSO Editors (2007), "Pi Puppids," *Meteor Showers Online*.
[526] FERAJ Editors (2020), "Pi-Puppids 1901-2100: activity predictions." *Feraj.ru*.
[527] In the isotope 24, 25 and 26 forms. Mark E. Lawler *et al.* (1989), "Iron, magnesium, and silicon in dust from Comet Halley," *Icarus* 80(2):225–242.

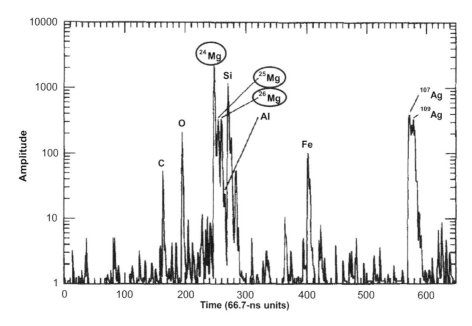

Figure 136: © Lawler
Spectrometer analysis of comet Halley

Comet Halley is not an isolated case. High concentrations of magnesium are found in most Jupiterian comets, *including Comet Grigg-Skjellerup*,[528] whose radar analysis[529] revealed high concentrations of olivine,[530] a magnesium-rich[531] mineral:

Size	ice	olivine	magnetite iron sulfide
mm	$3 \cdot 10^{17}$	$1.2 \cdot 10^{17}$	$6 \cdot 10^{16}$
cm	10^{11}	$4.5 \cdot 10^{10}$	$2 \cdot 10^{10}$

Figure 137: © Kamoun
Composition of Comet Grigg-Skjellerup

As shown above, the spikes in sulfur, nitrate and magnesium characterizing the December 1808 event might very well be due the overhead explosion of one, or several, cometary fragments associated with Comet 1808 C1 Grigg. Or it might be the result of a massive eruption or simultaneous eruptions. Both cometary and volcanic events leave similar traces.

In the end, separating cometary events from volcanic events doesn't really matter since comets can and do trigger volcanism and seismicity. This topic will be the main topic of the following chapter.

---

[528] Henner Busemann *et al.* (2009), "Ultra-primitive interplanetary dust particles from the comet 26P/Grigg–Skjellerup dust stream collection," *Earth and Planetary Science Letters* 288(1–2):44–57.
[529] Paul Kamoun (1983), "Radar Observations of Cometary Nuclei," Massachusetts Institute of Technology.
[530] William Simmons *et al.* (2019), "Olivine," *Encyclopædia Britannica*.
[531] Olivine contains 25.37% of magnesium mass wise.

Correlation between Cometary Activity and Volcanic Activity

So far we have suggested that some dust spikes caused by cometary events are probably wrongly attributed to volcanic eruptions. But volcanic activity and cometary events are not mutually exclusive. In the following, we will discover that there is a clear correlation between volcanism and cometary activity. Indeed, *cometary activity can be a direct cause of volcanic eruptions.*

Although seemingly radical to a modern mind, this idea is nothing new. In antiquity, philosophers considered comets as the trigger of volcanic eruptions and other calamities. For instance, English astronomer and philosopher Thomas Forster wrote the following words in 1829:

> The philosophers of antiquity almost universally believed the **approach of comet towards the sun to bring pestilence on the surface of the earth, by rousing volcanic fire**, and by disturbing the atmosphere.[532]

Forster was one of the last scientists to defend this ancient cosmology, which was progressively labeled as groundless superstition and thoroughly suppressed by modern science and its uniformitarian dogma where life on Earth is disconnected from cosmic events.

However, the facts are stubborn and the past few years have witnessed a re-emergence of this ancient 'heresy' with several papers acknowledging the correlation between, on one hand, solar activity and volcanic eruptions,[533] and on the other, solar activity and earthquakes.[534]

For example a recent publication discovered a "very high" correlation between large earthquakes and solar activity:

> We found clear correlation between proton density and the occurrence of large earthquakes (M>5.6), with a time shift of one day. **The significance of such correlation is very high, with probability to be wrong lower than 10^{-5}**. The correlation increases with the magnitude threshold of the seismic catalogue.[535]

---

[532] Thomas Forster (1829), *Illustrations of the atmospherical origin of epidemic disorders of health*, Meggy & Chalk, p. 14.

[533] Michele Casati (2014), "Significant statistically relationship between the great volcanic eruptions and the count of sunspots from 1610 to the present," *EGU General Assembly Conference Abstracts*. Dhani Herdiwijaya et al. (2014), "On the Relation between Solar and Global Volcanic Activities," *Proceedings of the 2014 International Conference on Physics*, pp. 105–108.

[534] I.P. Shestopalov et al. (2014), "Relationship between solar activity and global seismicity and neutrons of terrestrial origin," *Russian Journal of Earth Sciences* 14(ES1002). G. Duma et al. (2003), "Diurnal changes of earthquake activity and geomagnetic Sq-variations," *Natural Hazards and Earth System Sciences* 3(3/4):171–177. S. Odintsov et al. (2006), "Long-period trends in global seismic and geomagnetic activity and their relation to solar activity," *Phys. Chem. Earth* 31(1–3):88–93.

[535] V. Marchitelli et al. (2020), "On the correlation between solar activity and large earthquakes worldwide." *Scientific Reports* 10(11495).

Similarly the correlation between volcanic eruptions and solar activity appears to be very strong, especially for major eruptions:

> → Of the historical 31 large volcanic eruptions with index **VEI5+**, recorded between 1610 and 1955, **29 of these were recorded when the SSN**[536]**<46**. The remaining 2 eruptions were not recorded when the SSN<46, but rather **during solar maxima** of the solar cycle of the year 1739 and in the solar cycle No. 14 ...
> → Of the historical 8 large volcanic eruptions with index **VEI6+**, recorded from 1610 to the present, 7 of these were recorded with SSN<46 and more specifically, **within the three large solar minima known**: Maunder (1645–1710), Dalton (1790–1830) and during the solar minimums occurred between 1880 and 1920. As the only exception, we note the eruption of Pinatubo of June 1991, recorded in the **solar maximum** of cycle 22.
> → Of the historical 6 major volcanic eruptions with index VEI5+, recorded after 1955, 5 of these were not recorded during periods of low solar activity, but rather **during solar maxima**, of the cycles 19, 21 and 22.[537]

Extreme solar activity, either low (SSN<46) or high (solar maximum), seems to be associated with major volcanic eruptions, as illustrated by the triangles in the diagram on the next page.

Is there also a correlation between solar activity and earthquakes? According to Italian researcher Vito Marchitelli, there is indeed a high correlation:

> ... strongly statistically significant, evidence for a **high correlation between large worldwide earthquakes and the proton density near the magnetosphere, due to the solar wind**. This result is extremely important for seismological research and for possible future implications on earthquake forecast. In fact, although the non-poissonian character, and hence the correlation among large scale, worldwide earthquakes was known since several decades, this could be in principle explained by several mechanisms. In this paper, we demonstrate that it can likely be due to the **effect of solar wind, modulating the proton density and hence the electrical potential between the ionosphere and the Earth**.[538]

Solar activity influence is not, however, restricted to large earthquakes. The following summarizes the findings of several published papers and pinpoints magnetosphere-disrupting cosmic and solar radiations as a cause of *all* volcanic eruptions and earthquakes:

> The bottom line is that **all earthquakes and volcanic eruptions – big or small – are triggered by an external pressure induced on Earth's magnetic field. Strong Coronal Mass Ejection Flare directed at Earth can exert pressure that deform and shrink the Magnetosphere** by as much as 4 Earth Radius (4Re). But, the pressure would affect or impact the layers of Earth below its surface in different ways. It depends on the tectonics of each region. In some regions the tension would cause energy to be released in the form of earthquakes, while in others it would be in the form of volcanic eruptions.[539]

---

[536] SSN stands for Smoothed International Sunspot Number. A sunspot count lower than 46 suggests low solar activity.
[537] Casati (2014).
[538] V. Marchitelli *et al.* (2020), "On the correlation between solar activity and large earthquakes worldwide," *Scientific Reports* 10(11495).
[539] The Watchers Editors (2015), "Cosmic-solar radiation as the cause of earthquakes and volcanic eruptions," *The Watchers*.

Part III: Volcanoes, Earthquakes and the 3,600-Year Comet Cycle

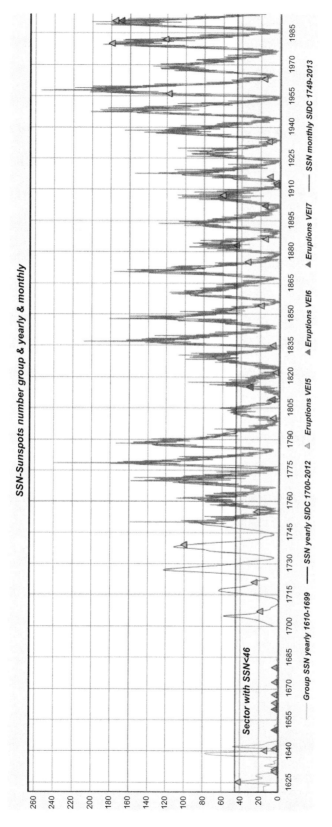

Figure 138: © Casati
Correlation between solar activity and major eruptions

The above shows that quakes and eruptions are similar phenomena in that they release tectonic energy. It also shows that sudden changes in solar activity, in particular CMEs,[540] deform the magnetosphere, which in turn affects the tectonics of our planet. What it *doesn't* do is provide any explanation about what modulates solar activity in the first place.

Interestingly, the presence of a comet in the solar system is correlated with spikes in solar activity like CMEs.[541] For mainstream science this correlation is a "simple coincidence."[542]

Is it just a coincidence though? Part of the answer was provided by Berkeley physicist Mensur Omerbashich[543] when he discovered a correlation between major earthquakes and cometary activity:

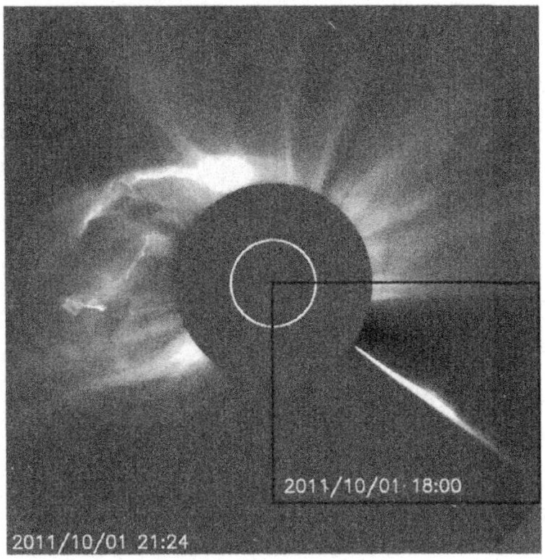

Figure 139: © SOHO/NASA/ESA
Massive coronal mass ejection shortly after a comet dove into the sun (inset, right).

> To add to my solution's robustness, I include alignments to the **comet C/2010 X1 (Elenin)** as it is the only heavenly body currently in our solar system besides planets, and show that **it impacted very strong seismicity since 2007** (and strongest seismicity, perhaps since 1965).[544]

The correlation between major earthquakes and alignment with Elenin is indeed robust. Of the 12 major earthquakes that occurred between April 2007 and March 2011, half of them – including the March 11th, 2011, M9.0 Japan earthquake – involved Elenin alignment, as shown in the table below, where Comet Elenin is highlighted by black rectangles:

ALIGNMENTS	DATE	MOON	LOCATION	MAGNITUDE	DEPTH	LABEL
Mars-Earth- Elenin ; **Moon**-Earth-**Sun**	Apr 01 2007	F	Solomon Isles	Mw=8.1	d=10.0 km	CMPS
Earth-Venus- Elenin	Aug 15 2007		Peru	Mw=8.0	d=39.0 km	CP
Earth-**Moon**-**Sun**	Sep 12 2007	N	Indonesia	Mw=8.5	d=30.0 km	MS
Earth-**Moon**-**Sun**	Sep 12 2007	N	Indonesia	Mw=7.9	d=30.0 km	MS
Earth-Moon-**Sun**	Dec 09 2007	N	Fiji	Mw=7.8	d=149.0 km	MS
Elenin -Earth-Neptune	May 12 2008		China	Mw=7.9	d=19.0 km	CP
Earth-Mercury-Jupiter	Jan 03 2009		Indonesia	Mw=7.7	d=17.0 km	P
Earth-Sun-Mercury	Jul 15 2009		New Zealand	Mw=7.8	d=12.0 km	PS
Earth-Venus- Elenin	Sep 29 2009		Samoa	Mw=8.1	d=18.0 km	CP
Venus-Earth-Uranus	Oct 07 2009		Vanuatu	Mw=7.8	d=35.0 km	P
Elenin -Earth-Sun; Earth-**Sun**-Jupiter; MS	Feb 27 2010	F	Chile	Mw=8.8	d=23.0 km	CMPS
Elenin-Earth-Sun; Earth-Mercury-Uranus	Mar 11 2011		Japan	Mw=9.0	d=32.0 km	CPS

Figure 140: © Omerbashich
Comparison of Earth's alignments vs. very strong earthquakes

---

[540] Coronal mass ejections.
[541] John W. Noonan *et al.* (2018), "Ultraviolet Observations of Coronal Mass Ejection Impact on Comet 67P/Churyumov by Rosetta Alice," *The Astronomical Journal* 156:16.
[542] Mike Wall (2011), "Did Comet Cause Solar Explosion? Hardly, Experts Say," *Space.com*.
[543] ORCID Editors (2020), "Mensur Omerbashich Biography," *Orcid.org*.
[544] Mensur Omerbashich (2012), "Astronomical alignments as the cause of ~M6+ seismicity," *arXiv*:1104.2036v7.

Beyond the proven correlations described above between comets and large earthquakes on one hand, and between comets and solar activity on the other, could there be *causation*, where cometary activity in the solar system *triggers* volcanism and seismicity?

To answer this question, here is a summary of the electrical mechanisms that enable comets to induce volcanic eruptions and earthquakes, via the modulation of solar activity. The following is extracted from my previous book,[545] which provides a much more detailed analysis of this topic.

The diagram below depicts the Sun, which is electrically active and positively charged. It is surrounded by an electrically negative layer called the heliosphere, as shown by the light grey ellipse, whose boundary (called the heliopause) extends beyond the solar system[546]:

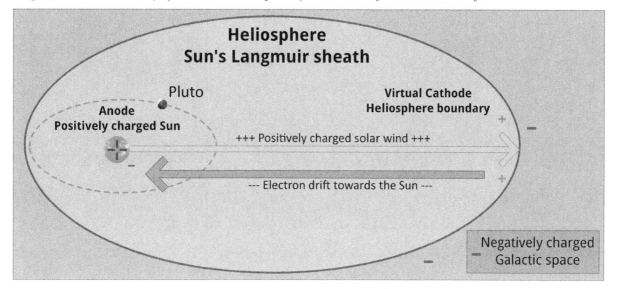

Figure 141: © Sott.net
The Sun and its heliosphere

The Sun–heliopause couple acts like a giant capacitor that is discharged by planetary alignments and/or foreign objects entering the solar system, in the same way a bug zapper discharges when a mosquito flies inside it.

A prime candidate for triggering such solar discharges are comets, because they are electrically very active, as indicated by their intense glow and because of their electrically conductive plasma tails that can extend hundreds of millions of kilometers.[547]

When a comet triggers a discharge of the solar 'capacitor,' the Sun releases solar flares, including coronal mass ejections (CMEs), which are massive quantities of protons – positively charged particles. These discharges, if properly oriented, can reach planet Earth.

---

[545] Lescaudron (2014), chapters 7 and 8.
[546] The Sun's heliopause – the external boundary of the heliosphere – is about 123 astronomical units (123 times the Sun–Earth distance) from the Sun. For comparison Pluto, the farthest planet in the solar system, is on average 'only' 40 AU from the Sun.
[547] The longest observed cometary tail belongs to Comet Hyakutake. In 1996, the Ulysse spacecraft measured its tail to be more than 500 million kilometers long, more than three times the distance between the Sun and Earth. See: G. Gloeckler *et al.* (2000), "Interception of comet Hyakutake's ion tail at a distance of 500 million kilometres," *Nature* 404:576–578.

The positively charged solar wind is depicted by the medium grey arrows on the left of the two diagrams below:

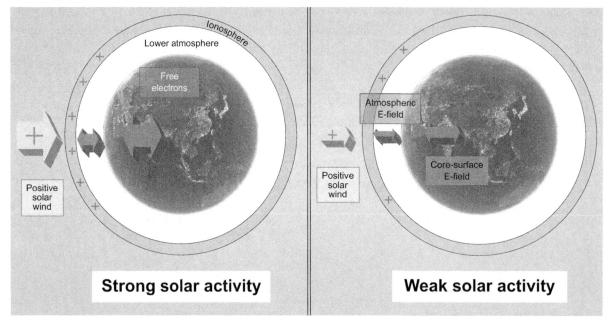

Figure 142: © Sott.net
Earth's electric fields and potentials according to solar activity

On the right of the image, solar activity is weak; therefore the Earth receives little positively charged solar wind. As a consequence, the electric potential of Earth's ionosphere is less positive and it tends to attract fewer free electrons from the Earth to its surface, making the surface less negatively charged. As a result, the electrical field between the ionosphere and the Earth's surface – 'atmospheric E-field,' represented by the medium grey double arrow – is reduced.

With fewer free electrons attracted from the inside of the Earth to its surface, the electric field between the Earth's surface and its core is also lowered, as shown by the white double arrow on the right picture.

This electric field is the binding force of our planet; it 'holds the planet together.' Extreme solar activity – like during solar minimum, solar maximum or major solar flares – can induce a sudden spike in the positive charge of the ionosphere, which translates into a sudden variation in the binding force. A crude but accurate enough analogy is if you were to hold an orange in your hand and then suddenly squeeze it or conversely holding it tight and suddenly releasing it.

The reduction in Earth's binding force also loosens the tectonic plates, which are then free to move relative to each other. It is this very movement – divergence, convergence or sliding – which is one of the main causes for earthquakes and volcanic eruptions.

This variation in Earth's binding force is not the only effect of cometary-triggered solar discharges. The Earth's spin is powered by the Sun too.[548] When the Earth is hit by a solar discharge, the spin rate experiences a minute change.

That's exactly what happened in August 1972 when an exceptional solar outburst hit the

---

[548]Lescaudron (2014), chapter 22.

Earth, provoking one such 'glitch': a sudden variation in the length of the day.[549] The electric disturbance from the solar flare affected the spin of the Earth such that, for a few days, it spun slower than usual. Because of its unusual power, the solar outburst flooded the Earth's double layer, diminishing its effectiveness and subsequently lowering the spin rate.[550]

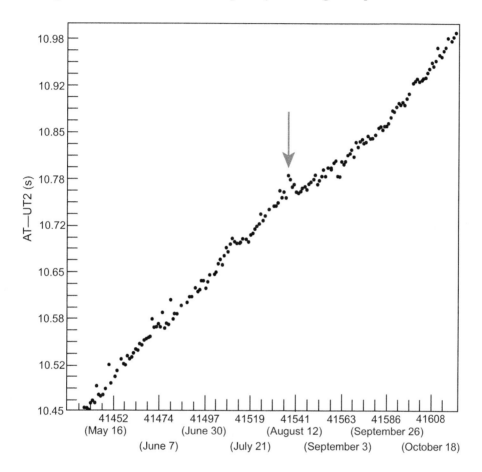

Figure 143: © Dachille
Change of slope in the day-length curve after the massive 1972 solar outburst.

Such changes in spin frequency can have two consequences:

1. *A minute crustal slippage.* The density of the crust is lower than the density of the mantle.[551] Therefore, the crust and the mantle won't slow down at the same rate. The mantle being denser, it has a higher momentum and won't slow down as fast as the crust. The difference in rotation between the crust and the mantle is equal to the crustal slippage. Crustal slippage, and the tremendous stress it exerts on the crust/mantle boundary, is a major cause of volcanism and seismicity:

---

[549] J. Gribbin *et al.* (1973), "Discontinuous Change in Earth's Spin Rate following Great Solar Storm of August 1972," *Nature* 243.
[550] A. De Grazia, and E. Milton (1984), *Solaria Binaria*, Metron Publications, p. 99.
[551] See Part I, chapter "Crustal slippage."

[Change] **in Earth's speed of rotation** would induce changes in the magma tide as it adjusted to the new equator or altered rotational speed. Such changes, however, might not be uniform throughout, owing to a 'drag' factor deep in the magma itself, although, overall, they would certainly **impose terrible strains on the lithosphere generally**.[552]

2. *A slight deformation of the shape of our planet.* On the left image of the diagram below, the planet is represented with a normal rate of rotation. Note that it's slightly broader at the level of the equator (ellipsoidal shape) because of the centrifugal force. On the right, Earth is depicted with a reduced rate of rotation. This slowdown induces a decrease in centrifugal force, which results in mechanical stress exerted on Earth's crust. Compression forces, as depicted by the white arrows, occur at low latitudes, while extension forces, depicted by the grey arrows, occur at higher latitudes. Thus the planet is deformed: its shape becomes less ellipsoidal and more spheroidal. Of course, this deformation has been exaggerated here (oval shape on the left vs. circular shape on the right) in order to make the effect more visually apparent:

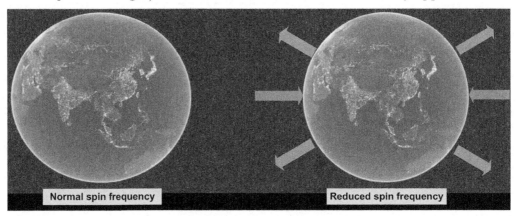

Figure 144: © Sott.net
How Earth spin rate affects its shape

The minute deformations of our planet caused by variations of solar activity induce tremendous mechanical stress on the Earth's crust. This deformation can even lead to partial ruptures around the weakest spots of the crust, i.e. the fault lines – boundaries between tectonic plates – which are the typical location of seismic and volcanic activity.

In summary, comets are major modulators of solar activity, which in turn affects the spin rate of our planet and its internal electric field. Both these phenomena can cause earthquakes and volcanic eruptions, through crustal slippage, crust deformation and reduction in the inner 'binding force.'

---

[552] D. Allan, and J. Delair (1997), *Cataclysm!*, Bear & Co., p. 180.

Conclusion

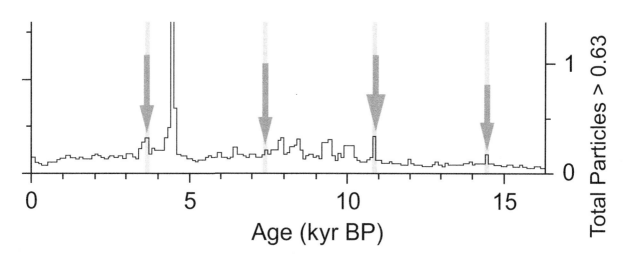

Figure 145: © Thomson
Dust concentration spikes in Huascaran core, ca. 14,400, 10,800, 7,200 and 3,600 BP

We have gathered data showing that Earth was probably subjected to cometary events – direct impact and/or overhead explosions and/or electric disruptions – in 14,400, 10,800, 7,200 and 3,600 BP. [553]

In light of these cometary-induced cyclical catastrophes, the fact that the Sumerians adopted as one of their main time units the 3,600 year (also known as '*Shar*') might not be a coincidence after all, but the reflection of a known cosmic cycle.

Keep in mind that each of the four earthly manifestations of the cometary cycle was different from the others. Each occurrence varied in location, magnitude, temperature drop, scope of extinction and other atmospheric variables. The table below recapitulates the features of each of the four events:

---

[553]The Huascaran graph also shows a major spike ca. 4,2000 BP, which marks was one of the most severe climatic events of the Holocene epoch. According to some researchers this event was due a cometary impact. See: Joachim Seifert and Frank Lemke (2019), "The Sumerian K8538 tablet. The great meteor impact devastating Mesopotamia – a 2019 translation addendum," *ResearchGate*.

Time (BP)	Location	Nature	Magnitude	Temperature drop	Extinctions	Other markers
14400	North / ice sheet impact ?	?	++++	About 10°	beginning of the late Quaternary megafaunal extinctions	SO4, CH4, CO2
10800	South - Oceanic impact?	Chondrite?	++	About 1°	End of numerous North american camps	NO3, Acetate, NH4, Oxalate, Formate, titanium, bromide, mercury, SO4
7200	South	Chondrite?	+++	A few degrees	End of numerous neolithic settlement	Mercury, SO4
3600	South	?	++	About 1°	End of the Xia dynasty	SO4

Figure 146: © Sott.net
3,600-year comet recapitulating table

We have also discovered that cometary events and volcanic eruptions leave very similar traces. In addition, they are not mutually exclusive: both can occur simultaneously for the simple reason that cometary events do cause volcanic eruptions.

Despite this proven causation, mainstream science continues to downplay the role of cometary events in favor of a systematic recourse to spontaneous (not comet-induced) volcanic eruptions to explain most catastrophes.

Oxford astrophysicist Victor Clube once wrote:

> Cynics would say that we do not need the celestial threat to disguise Cold War intentions; rather **we need the Cold War to disguise celestial intentions!**[554]

Keeping in mind the confusion between volcanic eruptions and cometary events combined with the prevalence of the volcanic hypothesis to explain sudden global cooling, we could paraphrase Clube as follows:

> Cynics would say that we do not need the celestial threat to disguise volcanic eruptions; rather **we need volcanic eruptions to disguise celestial intentions!**

And this disguise might be very timely. If a comet – or cometary swarm – interacted with the Earth ca. 14,400, 10,800, 7,200 and 3,600 BP, then we are due a repeat performance around now. Actually, the American Meteor Society's records of observed fireballs worldwide reveal a steady rise over the past 13 years,[555] suggesting that *the 'show' might actually have already begun*:

---

[554] Victor Clube, and Bill Napier (1990), *The Cosmic Winter*, B. Blackwell.

[555] With 386 observed events, 2020 is even higher the previous record of 2019 and its 327 events, although the number of observers decreased because of the numerous lockdowns and curfews. AMS Editors (2020), "Events stats," American Meteor Society Website. https://fireball.amsmeteors.org/members/imo_fireball_stats/

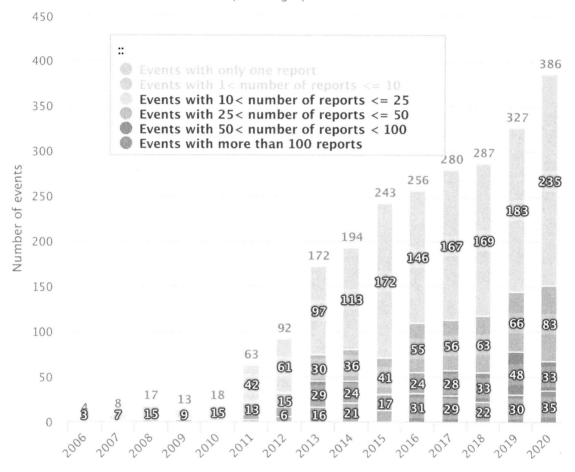

Figure 147: © AMS
Observed fireballs worldwide 2006–2020

Part IV:
The Seven Destructive Earth Passes of Comet Venus

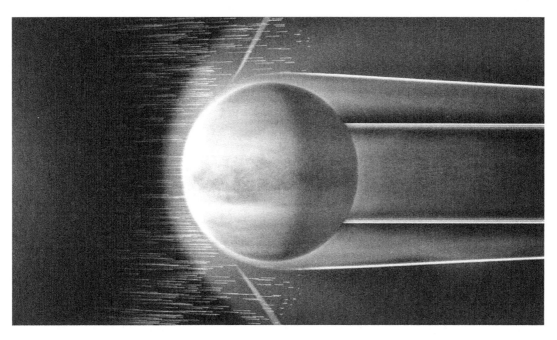

Figure 148: © NASA's Goddard Space Flight Center/G. Duberstein
Venus and its ion tail

Mars was knocked[556] *close to Earth by Venus, which was a cometary body at the time.* This radical idea was the topic of part II of this book, which focused mostly on the Earth–Mars interaction but left many questions concerning Venus unanswered.

What happened to Venus after it interacted with Mars? How long did it take for Venus to acquire its current, circular planetary orbit? Did cometary Venus have other interactions with planet Earth? How many passes did Venus make before it acquired a stable orbit? What were the dates of those passes? Did they have any effects?

Around 12,600 BP, cometary Venus was inside the solar system and knocked Mars towards Earth. Nowadays Venus is not a comet; it is a planet with a stable circular orbit.

Venus' status as a planet is attested as far back as the beginnings of the Sumer civilization ca. 4,500 BP.[557] This means that the transformation of Venus from comet to a planet occurred *between 12,600 BP and 4,500 BP*.

This transformation involved an orbital change: a transition from a long-duration highly elliptical orbit to a short-duration circular orbit. It is the *progressive* capture of cometary Venus by the Sun that transformed it into a stable planet, a process which probably involved several passes, with shorter and shorter orbital periods. This orbital change is similar to the one experienced by asteroids as depicted in the diagram to the right.

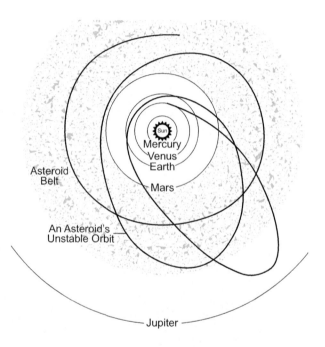

Figure 149: © Tufts University
Asteroid transition from an elliptical to a circular orbit

In order to determine when this progressive capture of Venus occurred, we must first identify the Venus markers in order to assess what kind of earthly parameters would have been modified by a fly-by of cometary Venus.

---

[556] Soon afterwards Mars, subjected to balancing gravitational forces reintegrated a stable quasi-circular orbit. For further explanations see: Rose *et al.* (1974).

[557] Paul Collins (1994), "The Sumerian Goddess Inanna (3400–2200 BC)," *Papers from the Institute of Archaeology* 5:103–118.

Venus Markers

Unlike Mars, *no meteorite yet found on Earth is known to be of Venusian origin,*[558] which suggests a limited transfer of solid material, if any. This scarcity is due to the fact that escape velocity on Venus is high[559] and elevated drag in the dense Venusian atmosphere would prevent anything reaching escape velocity and leaving the planet.[560]

So rather than look for rocks, we will focus on more volatile materials like the gases in Venus' atmosphere and tail that exist in higher concentrations than they do on Earth. Thus, a close encounter with Venus might be identified by *a sharp rise in terrestrial samples (pit cores, ice cores, etc.) of gases that are abundant on Venus.*

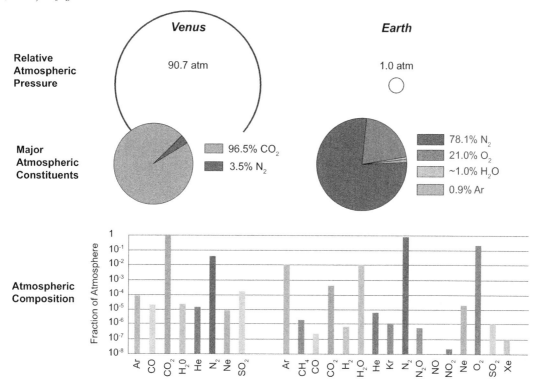

Figure 150: © Carie Franz
Venus atmosphere vs. Earth atmosphere

---

[558] NASA Editors (2019), "In Depth: Meteors & Meteorites," *NASA Science – Solar system exploration*.
[559] On Venus the escape velocity is 10.4 km/s. For comparison, it is only 6.5 km/s on Mars. B.J. Gladman *et al.* (1995), "Ejecta transfer between terrestrial planets," *Proceedings of the 172nd Symposium of the International Astronomical Union*.
[560] Mike Wall (2010), "Venus' Atmosphere Proves a Real Drag, Leading to a Discovery," *Space.com*.

The graph above shows the concentration of gases in the atmospheres of Venus and Earth respectively.

The diagram shows that two gases stand out as notably more concentrated on Venus than on Earth. These are carbon dioxide (CO_2 – as indicated by the light grey bars) and sulfur dioxide (SO_2 – as indicated by the medium grey). Notice that the scale is logarithmic, therefore a one gradation increase in the chart corresponds to a tenfold increase in terms of gas concentration.

CO_2 constitutes 96.5% of Venus' atmosphere and only about 400 ppm[561] of Earth's atmosphere,[562] which is a *2,500-fold difference*. As far as sulfur dioxide is concerned, Venus' atmosphere contains 186 ppm[563] while Earth's atmosphere only contains about 10 ppb,[564] which is a *20,000-fold difference*.

Beside those two gases, there is also deuterium, which is found on Earth at about 145 ppm,[565] while its presence on Venus is about 100 times higher.[566]

There are also hydrocarbon compounds, including one of the simplest forms – methane. Velikovsky hypothesized the presence of hydrocarbons in Venus' atmosphere as early as the 1950s.[567]

The idea that comets in general, and Venus in particular, contain hydrocarbons was ridiculed at the time by Carl Sagan and others. Three decades later, direct observations of Venus' high atmosphere proved Velikovsky right:

> "*Donahue*[568] *and his collaborators ... characterize the finding [of methane in Venus] as so surprising that they were loathe to publish them ...*"
>
> The researchers base their unlikely conclusion [that the methane is of volcanic origin] on the **abundance and composition of methane detected by a mass spectrometer aboard the Pioneer-Venus Probe**. Scientists had known for years that the spectrometer had recorded a sharp rise in methane, beginning at about 14 kilometers above the surface of Venus, during the probe's descent. But for nearly a decade, Donahue and his co-workers believed the surge merely reflected methane placed in the spectrometer on Earth in order to calibrate the instrument, not activity on Venus ...
>
> "*We concluded that the methane sampled was a primeval methane freshly vented from the planet's interior,*" says Donahue ... he estimates that a volcanic eruption spewing out **the amount of methane found by the Pioneer-Venus would occur only about once every 100 million years.**
>
> "*It is embarrassing to invoke such a wildly unlikely event as a chance encounter between the entry probe and a rare and geographically confined methane plume, but so far we have eliminated all other plausible explanations,*" Donahue added.[569]

---

[561] Parts per million, or 0.04%.

[562] Alan Buis (2019), "The Atmosphere: Getting a Handle on Carbon Dioxide," NASA Global Climate Change.

[563] V.I. Oyama *et al.* (1979), "Venus lower atmospheric composition: analysis by gas chromatography," *Science* 203(4382):802–805.

[564] Parts per billion. Weili Lin (2012), "Characteristics and recent trends of sulfur dioxide at urban, rural, and background sites in North China: Effectiveness of control measures," *Journal of Environmental Sciences* 24(1):34–49.

[565] M. Dörr (2011), "Deuterium," in *Encyclopedia of Astrobiology*, Springer.

[566] Catherine De Bergh *et al.* (1991), "Deuterium on Venus: Observations From Earth," *Science* 251(4993):547–549.

[567] Velikovsky (1950), chapter 2: "Naphtha."

[568] Thomas Donahue (1921–2004), American physicist, astronomer, and space scientist. He was involved in numerous space probe missions including the Pioneer Venus. See: Tamás Gombosil (2004), "Thomas Michael Donahue (1921–2004)," *Bulletin of the AAS* 36(4).

[569] Charles Ginenthal (2015), "Velikovsky's Hydrocarbon Clouds?," in *Carl Sagan and Immanuel Velikovsky*, Lulu Press.

Despite Velikovsky being proven correct, to maintain the mainstream scientific dogma that 'there is no methane in Venus' atmosphere' and 'therefore Velikovsky is wrong and uniformitarianism prevails,' scientists had to invoke an unsubstantiated and extremely unlikely cause to explain the methane in Venus' atmosphere:

> To explain the **large amount of methane found in the Venus atmosphere** the scientist said that the methane had to have come from an **extremely rare volcanic eruption**.
> The one explanation omitted by Donahue is that Venus has a good deal of methane in its atmosphere, just as Velikovsky predicted. Scientists would rather suggest a wildly improbable concept to explain the methane found than give consideration to Velikovsky's prediction.
> While **scientists like Sagan will call Velikovsky's theory extremely improbable, they will propose that it is probable that Pioneer-Venus just happened to come down on Venus to experience a unique event that happens once every hundred million years**.[570]

The presence of hydrocarbons around Venus as suggested by Velikovsky probably applies to Venus' 2-million-mile-long tail, because it contains carbon,[571] as detected by the *SOHO* probe, and hydrogen ions,[572] the two constituents of hydrocarbon. This applies in particular to methane,[573] which is the simplest form of hydrocarbon.

In total, we have identified four gases (SO_2, CO_2, D, CH_4) that are substantially more abundant in Venus' atmosphere than in Earth's atmosphere. A close encounter between the two celestial bodies should have left concentration spikes of those gases in earthly records.

Besides the gaseous spikes, cometary Venus should have left the typical signature of cometary encounters, i.e. increased dust due to:

- crossing the cometary tail,
- impact and/or overhead explosions of cometary fragments,
- induced volcanism/seismicity.[574]

This increase in atmospheric dust typically induces increased cloud cover, since dust particles act as a nucleation agent for cloud formation.[575] In turn, the increased cloud cover leads to *increased rainfall and a temperature drop*.

In total then, we have identified seven potential markers of a Venus encounter:

- Sulfur dioxide (SO_2)
- Carbon dioxide (CO_2)

Figure 151: © USSR Academy of Sciences
Venus' thick atmosphere and scorched surface taken by Russian probe Venera in 1981

---

[570] Ibid.
[571] Hecht (1997).
[572] Eduard Dubinin *et al.* (2013), "Plasma in the near Venus tail: Venus Express observations," *Journal of Geophysical Research: Space Physics* 118(12):7624–7634.
[573] Methane is made of one atom of carbon and four atoms of hydrogen, hence its formula: CH_4.
[574] As explained in part III, chapter "Correlation between Cometary Activity and Volcanic Activity."
[575] Lescaudron (2014), chapter 24.

- Deuterium (D)
- Methane (CH$_4$)
- Increased dust
- Increased precipitation
- Temperature drops

The diagram below recapitulates the hypothesized effects of nearby passes of comet Venus. The seven Venusian markers are shown in dark grey boxes:

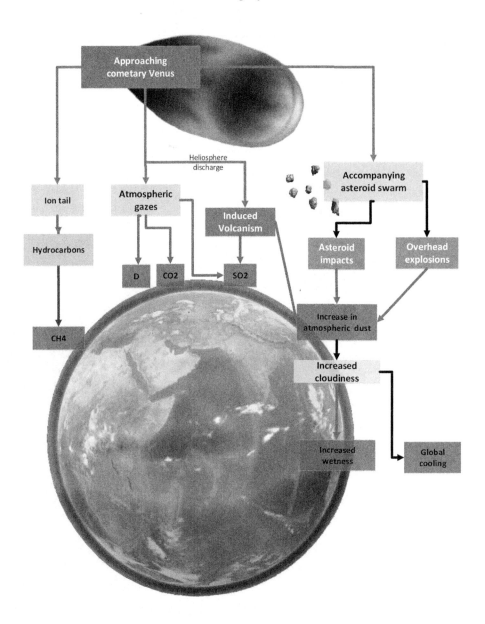

Figure 152: © Sott.net
Effects on Earth of close passes of cometary Venus

Looking for a Date

Now we are going to look at graphs of earthly records during the previously mentioned time range for Venus' close approach, i.e. between 12,600 and 4,500 BP. We will check if there is a date when the seven markers described above spike concomitantly.

Notice that we're not looking at every single pass of Venus yet, because most data don't provide a high enough resolution. Indeed, ice core analysis and the like usually come with a centennial or millennial scale, while Venus, especially during its last passes, must have exhibited an (almost circular) orbital period that is measured in decades.

Indeed, according to Velikovsky, the orbital period of cometary Venus was 52 years.[576] The period of a typical solar system comet (Jupiter-family comets) is less than 20 years,[577] any periodic comet has a period less than 200 years,[578] and the current orbital period of planetary Venus is only 224 days.[579]

So, we are looking for a span of *a few centuries* within which the hypothesized passes of Venus might have occurred.

---

[576] Velikovsky (1950), p. 161.
[577] Paul Weissman (2020), "Comet – The modern era," *Encyclopedia Britannica*.
[578] Ibid.
[579] Bruce Campbell *et al.* (2019), "The mean rotation rate of Venus from 29 years of Earth-based radar observations," *Icarus* 332:19–23.

Methane Spike and Temperature Drop

First, we are going to examine two indicators taken together: a methane spike in conjunction with a temperature drop. Because methane is a strong greenhouse gas, its global warming potential being 28 times that of carbon dioxide,[580] a methane spike should induce a *warming* not a cooling. Reciprocally, rising temperatures usually trigger a methane spike through heat-induced outgassing,[581] particularly from melting glaciers and melting ice caps.[582]

So, is there a date between 12,600 and 4,500 BP in which this unlikely conjunction – temperature drop and methane spike – occurred?

The charts below show temperature and methane records over the past 12,000 years. The part on the right with a grey background represents the 4,500 to 12,000 BP time span:

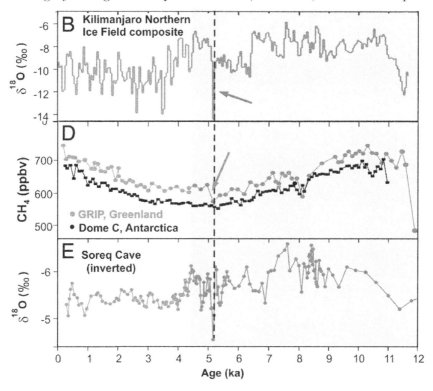

Figure 153: © Thomson
Temperature and CH$_4$ variations from 12,000 BP to now.

---

[580]Lancaster University Editors (2018), "Volcanoes and glaciers combine as powerful methane producers," *Phys.org*.
[581]S. Javadinejad *et al.* (2019), "Investigation of monthly and seasonal changes of methane gas with respect to climate change using satellite data," *Applied Water Science* 9(180).
[582]R. Burns *et al.* (2018), "Direct isotopic evidence of biogenic methane production and efflux from beneath a temperate glacier," *Scientific Reports* 8(17118).

Over the past 12,000 years, the largest temperature drop recorded in the Kilimanjaro Northern Ice Field, as indicated by the grey arrow in the top diagram, and in Israel's Soreq Cave, as shown by the grey arrow in the bottom diagram, *both occurred at the same time: 5,200 years BP (3,200 BC)*.

At the same date, *one of the largest methane spikes over the past 12,000 years* was recorded in the GRIP[583] Greenland ice core, as indicated by the black arrow in the middle diagram, with an increase from 600 to 650 ppbv.[584] Notice that the methane rise seems to last for a few centuries, which might suggest one lasting event or a series of events shortly interspersed.

Now, Kilimanjaro in Kenya and Soreq Cave in Israel are just two locations near the tropics. Was the 5,200 BP temperatures drop a local tropical glitch or a global event? The diagram below is a temperature reconstruction from the GISP2[585] ice core. The vertical black line marks the 5,200 BP date, which reveals a temperature drop as indicated by the light grey arrow:

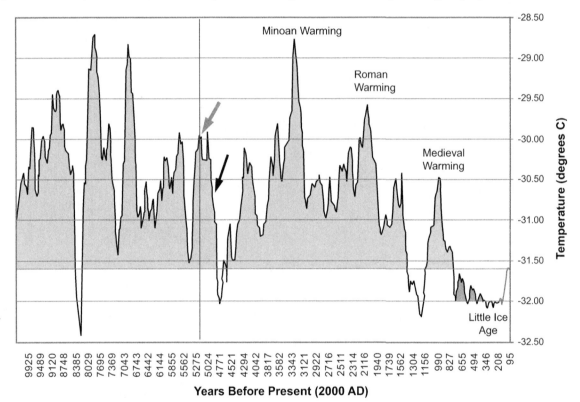

Figure 154: © Alley *et al.*
Greenland ice core - temperature over the last 10,000 years

The GISP2 reveals not only a temperature drop ca. 5,200 BP, but also a sustained and severe decrease in recorded temperatures, as shown by the black arrow, over the following centuries ca. 5,200–4,600 BP.

---

[583] Greenland Ice Core Project. European drilling project conducted in 1992.
[584] Part per billion per volume.
[585] Greenland Ice Sheet Project 2, conducted by an American team in 1993.

The marked cooling ca. 5,200 BP (3,200BC) is confirmed by dendrochronology[586] records from Ireland.

In the diagram on the next page, the 3,200 BC date corresponds to the bottom of the grey rectangle. It is characterized by a systematic spike in narrowness index,[587] as indicated by the black arrows.

According to dendrochronologist Mike Baillie, the 5,200 BP (3,200 BC) cooling event was *one of the three most severe our planet experienced over the past 7,000 years*:

> We ranked a very crude narrowness index, the product rs computed for a 10-yr window. We found that **the three highest values in the prehistoric period** [were] at 1153 BC, **3199 BC** and 4377 BC.[588]

Notice also that the cooling recorded ca. 5,200 BP seems to be followed by several centuries of cold episodes, as indicated by the succession of spikes in narrowness index in the grey rectangle, until ca. 4,600 BP, which corresponds to the top of the rectangle.

The cooling event in Ireland starting ca. 5,200 BP as revealed by dendrochronology was confirmed a few years later by a series of peat profile analyses.[589]

The above records coming from Kenya, Israel, Greenland and Ireland suggest that the 5,200 BP event started a lasting cooling episode that affected the whole planet. This period is known as the Piora Oscillation:

> The Piora Oscillation was an **abrupt cold and wet period** in the climate history of the Holocene Epoch; it is generally dated to the period of **c. 3200 to 2900 BCE**. Some researchers associate the Piora Oscillation with the end of the Atlantic climate regime, and the start of the Sub-Boreal, in the **Blytt-Sernander sequence of Holocene climates**.[590]

The Piora Oscillation got its name from the Piora Valley in Switzerland, where the 5,200 BP cooling event was first identified:

> ... some of the most dramatic evidence of the Piora Oscillation comes from the region of the Alps. Glaciers advanced in the Alps, apparently for the first time since the Holocene climatic optimum[591]; the **Alpine tree line dropped by 100 meters**.[592]

---

[586] Dendrochronology is the study of tree rings in order to reconstruct the climate of the past. See: Charles Ferguson (1970), "Concepts and Techniques of Dendrochronology," Laboratory of Tree-Ring Research, University of Arizona.

[587] In any given year, the narrower a tree ring is, the slower the tree growth was, and the colder the climate was. The narrower the ring, the higher the narrowness index is. See: L. Kairiukstis *et al.* (1986), *Methods of Dendrochronology – I. Proceedings of the Task Force Meeting on Methodology of Dendrochronology: East/West Approaches, Krakow, Poland*, IIASA/Polish Academy of Sciences, Warsaw.

[588] Mike Baillie *et al.* (1988), "Irish tree rings, Santorini and volcanic dust veils," *Nature* 332:344–346.

[589] C. Caseldine *et al.* (2005), "Evidence for an extreme climatic event on Achill Island, Co. Mayo, Ireland around 5200–5100 cal. yr BP," *Journal of Quaternary Science* 20(2):169–178.

[590] Wikipedia Editors (2007), "The Piora Oscillation," *Wikipedia*.

[591] Warm period during the 9,000 to 5,200 BP interval.

[592] Hubert F. Lamb (1995), *Climate, History, and the Modern World*, London: Routledge.

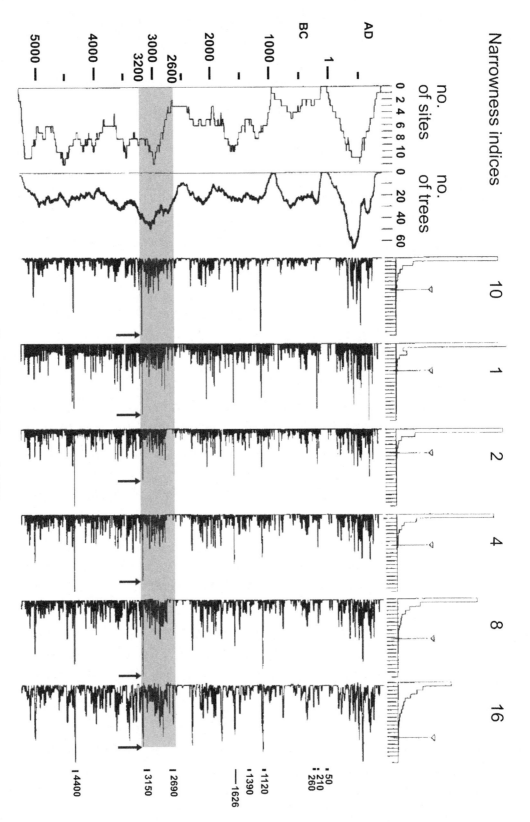

Figure 155: © Baillie et al., 1988
Tree rings narrowness index (Irish oaks)

Figure 156: © Ticino
Current tree line in the Piora Valley

So far we have found one date, 5,200 BP, that shows the unusual conjunction between a methane spike and marked a drop in global temperatures. Let's focus now on the five other Venusian markers and see if they display any incongruity ca. 5,200 BP.

Deuterium

Deuterium is an isotope of hydrogen – also known as heavy hydrogen.[593] As shown in the picture to the right, its nucleus constitutes one proton and one neutron, and its symbol is ^{2}H or D.

Deuterium is one of the most thoroughly measured chemicals of Venus because it is related to the presence of water, considered one of the necessary components for life to develop.[594] Venusian life or not, deuterium is far more abundant in Venus' atmosphere than on Earth:

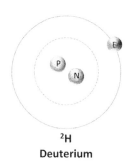

Figure 157: © Shala Howell Deuterium molecule

> Absorption lines of HDO and H$_2$O have been detected in a 0.23-wave number resolution spectrum of the dark side of Venus in the interval 2.34 to 2.43 micrometers, where the atmosphere is sounded in the altitude range from 32 to 42 kilometers (8 to 3 bars). **The resulting value of the deuterium-to-hydrogen ratio (D/H) is 120 ± 40 times**[595] **the telluric ratio**, providing unequivocal confirmation of in situ Pioneer Venus mass spectrometer measurements that were in apparent conflict with an upper limit set from International Ultraviolet Explorer spectra. **The 100-fold enrichment of the D/H ratio on Venus compared to Earth** is thus a fundamental constraint on models for its atmospheric evolution.[596]

Not only was a high concentration of deuterium found in Venus' high atmosphere, but this deuterium is pushed by solar winds outside the Venusian atmosphere towards Venus' surrounding space and ion tail:

> ... the motional electric field of the solar wind impressed across the draped magnetic field lines of the ionotail of Venus eventually overtakes the polarization electric field and accelerates the ions up to solar wind speeds as the ionotail merges into the interplanetary medium. **Essentially all escape of H* and D* by the electric field process occurs in the light ion bulge, where most of these ions reside.**[597]

This leakage of deuterium from Venus' high atmosphere to its surrounding space and ion tail makes a theorized gas transfer from Venus to Earth more likely if the two bodies were close enough.

---

[593] William Lee Jolly (1999), "Hydrogen chemical element," *Encyclopedia Britannica*.

[594] J. Bertaux et al. (1989), "Deuterium content of the Venus atmosphere," *Nature* 338:567–568.

[595] Donahue found even higher ratios of deuterium to normal hydrogen on Venus compared to Earth: 150 ± 30 or 157 ± 30 or 138 times. See: T. Donahue et al. (1997), "Ion/neutral Escape of Hydrogen and Deuterium: Evolution of Water," in *Venus II: Geology, Geophysics, Atmosphere, and Solar Wind Environment*, University of Arizona Press, p. 385.

[596] De Bergh et al. (1991).

[597] E. Dubinin et al. (2017), "The effect of solar wind variations on the escape of oxygen ions from Mars through different channels: MAVEN observations," *Journal of Geophysical Research: Space Physics* 122:11,285–11,301.

Earlier in this book,[598] we emphasized that even today, as a stable planet, Venus retains a very long ion tail. Venus' tail is 45 million km[599] long, so long in fact that its ion tail almost reaches Earth[600] when the Sun, Venus and Earth are aligned, as shown in the diagram below:

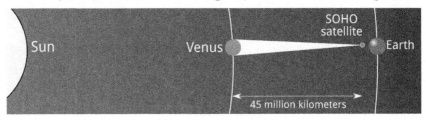

Figure 158: © New Scientist
Venus' ion tail

It's probable that when Venus was a comet, its ion tail was much larger, hundreds of millions of km long, making the transfer of ions – including deuterium ions – from its tail to Earth possible, even if the two bodies were at a considerable distance from each other.

Now let's look at deuterium records on Earth. Since deuterium concentration fluctuates a lot, we'll focus on the deuterium excess that helps identify spikes more easily:

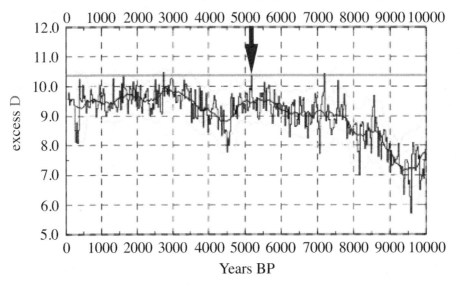

Figure 159: © Masson-Delmotte
Deuterium excess over the past 10,000 years

The above chart shows the deuterium excess found in Greenland ice cores (GRIP). We can see that ca. 5,200 BP, as indicated by the black vertical arrow, our planet experienced *one of the three largest deuterium spikes of the past 10,000 years*, reaching 10.4 excess deuterium, as shown by the light grey horizontal line.

---

[598] Part II, chapter "Was Venus a Comet?"
[599] 29 million miles.
[600] Hecht (1997).

Sulfur Dioxide

As mentioned previously, Venus' atmosphere contains 186 ppm of SO_2 while Earth's atmosphere only contains about 10 ppb – a 20,000-fold difference.

For comparison, the 1883 Krakatoa eruption,[601] one of the largest eruptions in modern history, generated a 40 ppm spike[602] in the GISP2 Greenland ice core.

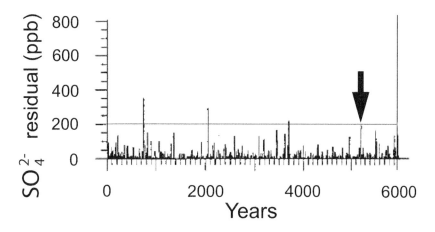

Figure 160: Zelinsky
Greenland SO_2 residual concentration over the past 6,000 years

In the diagram above, the black arrow shows a 200-ppm residual sulfur dioxide spike ca. 5,200 BP (3,200 BC). This is the *fourth-strongest sulfur signal recorded over the past 6,000 years*.

The 5,200 BP event generated 200 ppm of residual SO_2 and 255 ppm[603] of total SO_2, which is more than six times larger than the Krakatoa sulfur signal.

As discussed earlier,[604] mainstream science considers SO_2 spikes as solely due to volcanic eruptions. Interestingly, the 5,200 BP signal is considered as *the largest unidentified spike in the last five millennia*, as shown by the black arrow with the large question marks in the diagram below:

---

[601] For more details about the Krakatoa eruption, see part III, chapter "The 3600 BP (1600 BC) Event."
[602] Oppenheimer (2003).
[603] C. Hammer *et al.* (1980), "W. Greenland ice sheet evidence of post-glacial volcanism and its climatic impact," *Nature* 288:230–235.
[604] See part III, chapter "Mystery Eruption."

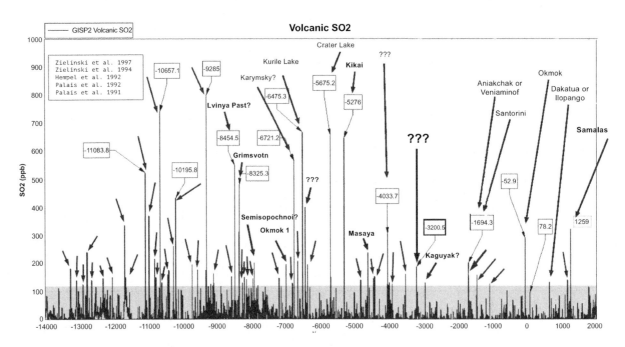

Figure 161: © Volcano café
GISP ice core SO$_2$ concentration over the past 16,000 years

The massive 5,200 BP sulfur spike has been tentatively associated with the eruption of the Avachinsky volcano in Kamchatka. This eruption occurred in 5,200 +/- 150 BP, so the timing might fit, but the Volcanic Activity Index[605] (VEI) for this eruption was only 5, and it released a mere 1.1 km^3 of tephra.[606]

For comparison, the 1883 Krakatoa eruption had a VEI of 6 and released almost 20 times more material, with 21 km^3 of tephra.[607] Although, as previously mentioned, the Krakatoa sulfur release was less than 1/6th of the 5,200 BP spike.

The Avachinsky eruption might have been a small contributor to the 5,200 BP sulfur spike, but it is very unlikely to have been its main cause.

---

[605] The VEI reflects how much volcanic material is thrown out, to what height, and how long the eruption lasts. It is a logarithmic scale, so an increase of 1 index indicates an eruption that is 10 times as powerful. See: Christopher Newhall *et al.* (1982), "The Volcanic Explosivity Index (VEI): An Estimate of Explosive Magnitude for Historical Volcanism," *Journal of Geophysical Research* 87(C2):1231–1238.

[606] 0.26 cubic miles. Tephra is rock fragments and particles ejected by a volcanic eruption. E. Venzke (2013), "Global Volcanism Program, volcanoes of the World," Smithsonian Institution.

[607] 5 cubic miles. Deborah Hopkinson (2004), "The Volcano That Shook the world: Krakatoa 1883," *Scholastic.com Storyworks* 11(4):8.

Carbon Dioxide

CO_2 is by far the most prevalent gas in Venus' atmosphere (96.5%), whereas it is only a trace gas in Earth's (400 ppm). If there were passes of cometary Venus near Earth ca. 5,200 BP, one could expect a gaseous transfer and a subsequent CO_2 spike in Earth records.

This is indeed what CO_2 concentration measured in the EPICA[608] ice core suggests:

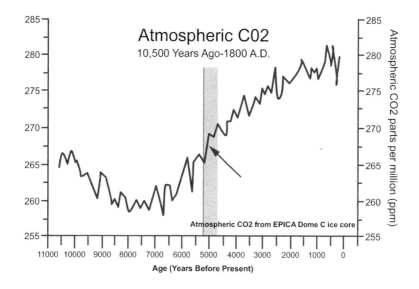

Figure 162: © Rosenthal
CO2 concentration in EPICA ice core (11,000 BP–now)

Notice that, as with the previously mentioned global cooling and methane spike, the increase in atmospheric CO_2 lasted several centuries. As depicted by the grey arrow in the diagram above, CO_2 concentration rose from ca. 5,200 BP, as shown by the vertical grey line. This rise lasted *about six centuries* until ca. 4,600 BP, as shown by the light grey rectangle[609].

This lasting rise in atmospheric CO_2 is all the more unexpected given that it is usually associated with warming periods and the induced massive release of CO_2 by warming oceans.[610] As we now know, the 5,200–4,600 BP period was not a warming but a global *cooling* episode, which should have increased the absorption of CO_2 by ocean and consequently dropped the concentration in atmospheric CO_2. The diagram above shows that just the opposite happened, giving extra credence to an intake of extraterrestrial CO_2.

---

[608] European Project for Ice Coring in Antarctica. Deep drilling project started in 1996.
[609] The spike identified in the above graph appears to be part of a larger trend of rising CO2. It seems that this long lasting increase is due to oceanic CO2 release caused by the global warming that marked most of the Holocene. See: Ward, Dale (2016), "The climate of the Holocene," *University of Arizona*.
[610] Wendy Zukerman (2011), "Warmer oceans release CO2 faster than thought," *New Scientist*.

Increased Atmospheric Dust

Cometary events, like the hypothesized close approach of cometary Venus, can lead to direct impacts, overhead atmospheric explosions and the crossing of Venus' cometary tail. In addition, we now know that the proximity of a massive cometary body like Venus can induce earthquakes and volcanic eruptions.

These four, non-mutually exclusive, features are *all potential causes for an increase in atmospheric dust*.

Dust analysis conducted in Terra Del Fuego, Argentina, reveals a moderate dust spike ca. 5,200 BP,[611] as indicated by the vertical black line and the medium grey arrow in the diagram below. This increase in dust concentration lasted a few centuries up until ca. 4,600 BP, as shown by the medium grey rectangle:

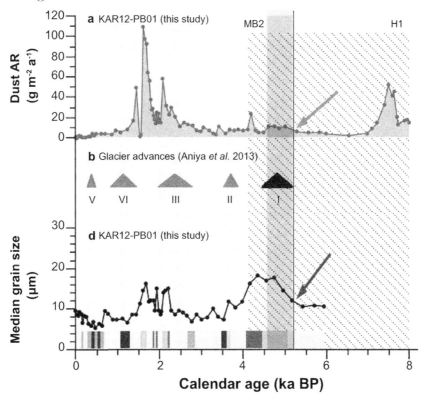

Figure 163: © Vanneste
Dust concentration and size, glacier advance (8,000 BP–now)

---

[611]H. Vanneste *et al.* (2016), "Elevated dust deposition in Tierra del Fuego (Chile) resulting from Neoglacial Darwin Cordillera glacier fluctuations," *Journal of Quaternary Science* 31:713–722.

The diagram above also reveals that 5,200 BP marks the beginning of a period of glacier advance that lasted until ca. 4,600 BP, as shown by the black triangle, confirming the global cooling described above.

Notice also that, as shown by the dark grey arrow, median dust grain size increased markedly ca. 5,200 BP, suggesting that the main dust source was not the usual winds blowing over an arid region, because this eolian phenomenon tends to carry small dust particles.[612] This means that unusually strong winds and/or extraterrestrial dust were probably involved in this dust deposit.

Tierra Del Fuego is not the only place that witnessed an increase in atmospheric dust at this time. In Huascaran, Peru, ice cores reveal a similar pattern, with a marked dust spike.[613] The diagram below exhibits a two-fold increase in dust concentration, ca. 5,200 BP, as shown by the vertical black line. This dust spike is the sixth largest in 10,000 years and it lasted until ca. 4,600 BP, as depicted by the light grey rectangle.

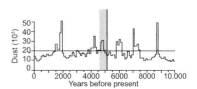

Figure 164: © Thomson
Huascaran dust concentration
(10,000 BP–now)

Thousands of kilometers from Peru and Argentina, the African continent also experienced dust spikes. As shown in the diagram below, both the Kilimanjaro ice core and the Gulf of Oman eolian deposition reveal a dust spike starting ca. 5,200 BP, as shown by the black vertical line, and lasting until ca. 4,600 BP, as indicated by the light gray box:

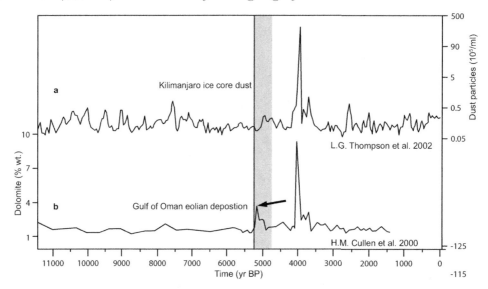

Figure 165: © Thomson et al., 2002 – Cullen et al., 2000
Dust concentration in Kilimanjaro and Gulf of Oman

Notice also that in the diagram above, the dust spike in the Gulf of Oman, as indicated by the black arrow, was the second largest in the past 11,000 years.

---

[612] J-B.W. Stuut et al. (2014), "The significance of particle size of long-range transported mineral dust," *PAGES News* 22(2):14–15.

[613] L.G. Thompson et al. (1995), "Late Glacial Stage and Holocene Tropical Ice Core Records from Huascarán, Peru," *Science* 46-50.

Increased Wetness

Atmospheric dust particles act as condensation nuclei and stimulate cloud formation.[614] Therefore, the above-mentioned increase in atmospheric dust ca. 5,200 BP likely led to an increase in cloudiness. As a general rule, increased cloudiness results in increased rainfall,[615] as depicted in the drawing below:

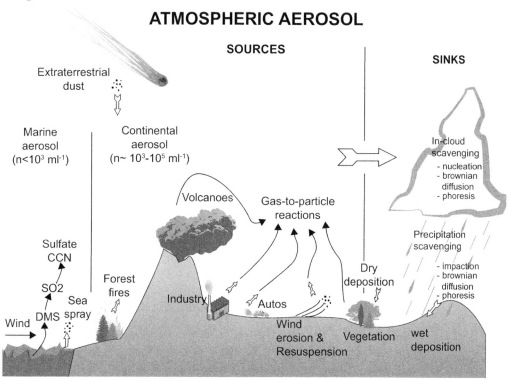

Figure 166: © NOAA
Atmospheric dust influence on precipitation

Nevertheless, an increase in cloudiness doesn't *always* lead to an increase in rainfall. As shown by Vandana Jha, increase in atmospheric dust tends to increase precipitation in *wet* systems while reducing precipitation in *dry* systems.[616] From this perspective, atmospheric dust can be seen as a catalyst of pre-existing weather condition, making wet areas wetter and dry areas drier. As

---

[614] Stephen Cole (2007), "Global 'Sunscreen' Has Likely Thinned, Report NASA Scientists," NASA.

[615] V.-M. Kerminen *et al.* (2012), "Cloud condensation nuclei production associated with atmospheric nucleation: a synthesis based on existing literature and new results," *Atmospheric Chemistry and Physics* 12(24):12037–12059.

[616] Vandana Jha (2018), "Sensitivity Studies on the Impact of Dust and Aerosol Pollution Acting as Cloud Nucleating Aerosol on Orographic Precipitation in the Colorado River Basin," *Advances in Meteorology*.

we will now see, studies of the changes in precipitation ca. 5,200 BP depict a similarly nuanced picture.

The diagram below shows rainfall over the past 10,000 years in the Indus Valley.[617] As shown by the black arrow and the black vertical line, the 5,200 BP date reveals *a sharp increase in rainfall* from 450 mm to about 800 mm a year,[618] roughly an 80% increase. In addition this increase in precipitation lasted for several centuries after 5,200 BP, as shown by the light grey box:

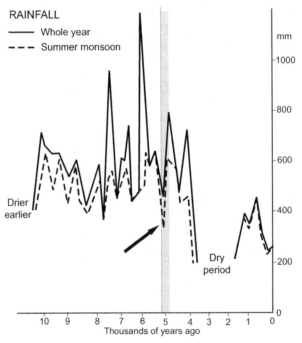

Figure 167: © Lamb *et al.*, 1978
Rainfall over the past 10,000 years in the Indus Valley

Thousands of kilometers away from the Indus Valley, increased wetness at this time is confirmed by research conducted around the Dead Sea. The salt caves of Mount Seldom are 100 meters[619] above sea level. Twigs and leaves from an oak tree[620] were found in one of them, preserved over millennia. How do we explain the oddity of oak fragments in a sterile salt cave 100 meters above sea level?

> Now it is geologically certain that it was an ancient **pluvial age that created caves in the salt**; in fact past climate can be inferred by carefully measuring the width of caves formed by salt dissolution. The cave widths can in turn be compared with correlative glacial advances in northern Europe (bigger caves = more rain = more glaciers) and the cave elevations with ancient sea levels of the Dead Sea itself.
>
> The horizon of wide caves found some 300 feet above the present sea level necessarily indicates

---

[617] Lamb (1995).
[618] From 18 to 31 inches.
[619] 300 feet.
[620] *Quercus Calliprinus* – Israeli Oak.

an **extremely wet period in the early Bronze Age, or about 4200 to 5200 radiocarbon years before present**.

Oak twigs, driftwood, and marl found in the caves **must have been transported by floodwater** from some other part of the Judiah Hills, when the **water level was some 300 feet higher than present, implying heavy flooding** on the Jordan River and coupled probably with lower evaporation rates due to cooler weather.[621]

The diagram below illustrates the increased wetness in the Dead Sea region experienced ca. 5,200 BP, as shown by the black vertical line. Notice that the wet episode around the Dead Sea lasted several centuries, as indicated by the medium grey box, the end of which occurred between 4,900 and 4,400 BP. Indeed during those 5 centuries the sea level reconstruction is uncertain, as shown by the question marks and the dashed sea level curve:

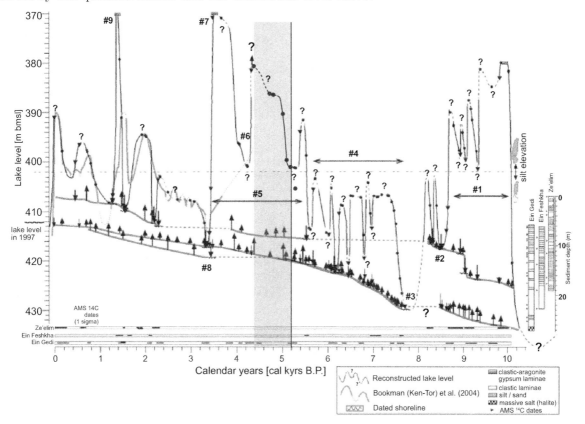

Figure 168: © Migowski
Dead Sea level over the past 10,000 years

Geologist Harry Rowe discovered a similar phenomenon on the American continent where the Lago Grande[622] started to rise ca. 5,200 BP.[623] The rise lasted for centuries until the lake was

---

[621] Richard Meehan (1999), "Oak," in *Ignatius Donnelly and the end of the world*, Kirribili Press.
[622] The Lago Grande is the main part of Lake Titicaca. It is situated along the border between Bolivia and Peru.
[623] Harry Rowe *et al.* (2004), "Hydrologic-energy balance constraints on the Holocene lake-level history of lake Titicaca, South America," *Climate Dynamics* 23:439–454.

about 100 meters[624] higher, which, coincidentally or not, is the same figure we encountered for the Dead Sea rise.

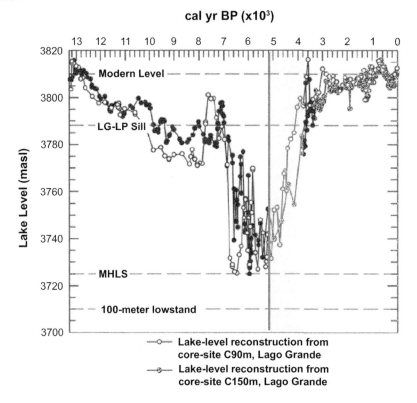

Figure 169: © Harry Rowe
Lago Grande level over the past 13,000 years

The water rise in Asia (Indus Valley), the Middle East (Dead Sea) and the Americas (Lake Titicaca) strongly suggests that our planet experienced a marked wet episode starting ca. 5,200 BP and lasting for several centuries.

Notice however that while numerous areas experienced some increased wetness between 5,200 and 4,600 BP, some other areas experienced an aridification, as was the case for Spain, for example:

> Palynological data from areas within the thermo- and mesomediterranean areas reported **woodland cover reductions after ca. 5,200 BP**. ...
>
> During this period, an increase in fire activity, probably enhanced by arid climate conditions, may have played a crucial role in favoring the spread of sclerophyte and fire-prone communities. ...
>
> In addition, **marked changes in several lake sequences took place approximately at 5,100 BP**. In Villarquemado, deposition in ephemeral lake conditions continued without major changes in the geochemical signature (SUB-2A), except for a significant increase in Mn that might reflect **higher occurrence of oxidation processes in a shallow environment**.
>
> Other pollen-independent studies reach similar conclusions: at Laguna de Medina, Reed et al. (2001) suggest a clear decrease in lake levels after 5530 cal yr BP, while **at Siles phases of dramatic lake desiccation around 5200** and 4100 BP are identified.[625]

---

[624] 300 feet.

On the other side of the Mediterranean Sea, in North Africa, a similar aridification took place:

Arid interval 5010-4860 (+/- 150) at Tigalmamine in montane Morocco. Corresponding decline in oaks (*Quercus rotundifolia* and *canariensis*) in favor of Gramineae suggests **reduced winter precipitation** corresponding to cooler sea temperatures in North Atlantic.[626]

The above suggests that our planet experienced a dramatic and centuries-long wet period starting ca. 5,200 BP, *except southwest Europe and North Africa*, which experienced aridification.

In summary, so far, we have identified seven potential Venusian markers, namely temperature drop, methane, deuterium, sulfur dioxide, carbon dioxide, wetness and atmospheric dust. We have discovered that *there's only one date between 12,600 BP and 4,500 BP that stacks up with each of those seven markers: 5,200 BP*. This concomitant spike for each of the seven Venusian markers suggests that the first close approach of cometary Venus occurred 5,200 years ago.

In addition, most of those seven markers reveal a disruption that started ca. 5,200 BP and lasted until about 4,600 BP, which gives credence to several close passes of cometary Venus during this six-century time frame.

---

[625] Nick Brooks (2012), "Beyond collapse: Climate change and causality during the Middle Holocene Climatic Transition, 6400–5000 years before present," *Geografisk Tidsskrift-Danish Journal of Geography* 112:93–104.

[626] H.F. Lamb *et al.* (1995), "Relation between century-scale Holocene and intervals in tropical and temperate zones," *Nature* 373:134.

Impact on Human Populations

The Neolithic age ended around 5,000 BP and was superseded by the Bronze Age. Thus the period we are studying (5,200–4,600 BP) is situated across the late Neolithic Age and the early Bronze Age.

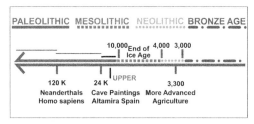

Figure 170: © Sott.net
Time periods (120,000 to 3,000 BC)

Despite the relative scarcity of archeological data for these remote times, we can see that *several cultures and settlements collapsed between 5,200 and 4,600 BP* during the Earth changes described above.

In Mesopotamia, the collapse of the Uruk culture after flourishing for 6 centuries has been pinpointed to around 5,200 BP,[627] as Uruk 'colonies' in the north were abandoned. Some smaller settlements in southern Mesopotamia were abandoned too. According to Italian geologist Peter Martini,[628] the collapse of the Uruk culture was due to *rapid cooling*.

Figure 171: © Benjamin Rabe
Cone mosaics covering a wall in Uruk, Iraq

---

[627] Brooks (2012).
[628] Peter Martini *et al.* (2010), *Landscapes and Societies: Selected Cases*, Springer Science & Business Media.

Another example from Mesopotamia is Jemdet Nasr,[629] a settlement mound in Iraq encompassing 5 hectares.[630] After two centuries of development, a rapid collapse[631] occurred at 4,900 BP. Irrigation cultures continued in the region, but new strong dynasties thrived again in Mesopotamia *only after 4,600 BP*, after an increase of temperatures and higher precipitation.

In the Nile valley, after centuries of an increase in population and social complexity, 5,200 BP marked the collapse of northern Egypt, which was subjugated by southern Egypt.[632]

Like in Mesopotamia, the collapse of northern Egypt is explained by "climatic deterioration":

> This occurred against a background of population agglomeration in the Nile Valley at a **time of increasing climatic and environmental deterioration** and uncertainty, which may have played an important role in driving competition over resources.[633]

According to French archeologist Robert Vernet, the Sahara region experienced a similar collapse with a sudden and marked reduction in the number of occupation sites north of 23°N over the 5200–5000 BP period.[634] The chart below shows this decrease in human population ca. 5,000 BP (black vertical line) north of 23° latitude (blue columns). The human population *decreased by about 50%* after more than two millennia of relative stability.

Notice that south of 23° latitude, the population decreased too (dashed line), but in a less pronounced way:

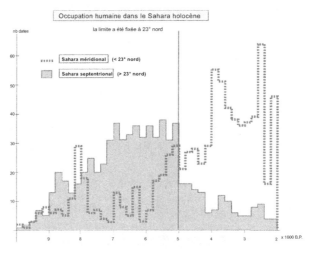

Figure 172: © Vernet *et al.*, 2000
Human occupation of the Sahara

The Knap of Howar on the island of Papa Westray in Orkney, Scotland, is a Neolithic settlement

---

[629] S. Pollock (1992), "Bureaucrats and managers, peasants and pastoralists, imperialists and traders: Research on the Uruk and Jemdet Nasr periods in Mesopotamia," *Journal of World Prehistory* 6:297–336.
[630] 14 acres.
[631] Joachim Seifert and Frank Lemke (201b6), "Climate Pattern Recognition In The Mid-To-Late Holocene (2900 BC To 1650 BC, Part 4)," *ResearchGate*.
[632] Brooks (2012).
[633] Ibid.
[634] Robert Vernet *et al.* (2000), "Isotopic Chronology of the Sahara and the Sahel During the Late Pleistocene and the Early and Mid-Holocene (15,000–6000 BP)," *Quaternary International* 68:385–387.

which may be the oldest stone building in northern Europe. Radiocarbon dating shows that it was deserted ca. 4,800 BP after nine centuries of continuous occupation.[635]

Figure 173: © Message to Eagle
The Knap of Howar, one of the oldest Neolithic complexes, Orkney, Scotland

The Eastern European Cucuteni-Trypillia culture emerged ca. 5,800 BP and flourished for six centuries. It encompassed hundreds of square kilometers, about 3,000 structures[636] and had tens of thousands of inhabitants.[637]

It eventually came to an abrupt end ca. 5,200 BP. For a long time the collapse of the Cucuteni-Trypillia culture was attributed to invasions.[638] But recently the climate change explanation has gained traction:

> In the 1990s and 2000s, another theory regarding the end of the Cucuteni-Trypillia culture emerged **based on climatic change** that took place at the end of their culture's existence that is known as the Blytt-Sernander Sub-Boreal phase. **Beginning around 3200 BC, the earth's climate became colder and drier than it had ever been since the end of the last Ice age, resulting in the worst drought in the history of Europe since the beginning of agriculture.**[639]

In northern China, the Yangshao culture bloomed for two millennia. The typical Yangshao village covered four to six hectares[640] and comprised a dozen houses made of clay and straw surrounded by a moat,[641] as shown by the picture below.

---

[635] Message To Eagle Editors (2017), "The Knap Of Howar: One Of The Oldest And Well-Preserved Neolithic Complexes Orkney, Scotland," *Message To Eagle*.
[636] Aleksandr Diachenko *et al.* (2012), "The gravity model: monitoring the formation and development of the Tripolye culture giant-settlements in Ukraine," *Journal of Archaeological Science* 39(8):2810–2817.
[637] Johannes Müller *et al.* (2016), *Trypillia Mega-Sites and European Prehistory: 4100–3400 BCE*, Taylor & Francis, p. 347.
[638] Marija Gimbutas (1956), "The prehistory of Eastern Europe, pt. 1: Mesolithic, Neolithic and copper age cultures in Russia and the Baltic area," *American School of Prehistoric Research*.
[639] David W. Anthony (2010), *The Horse, the Wheel, and Language: How Bronze-Age Riders from the Eurasian Steppes Shaped the Modern World*, Princeton University Press, p. 164–190.
[640] Ten to fourteen acres.
[641] Facts and Details Editors (2016), "Yangshao Culture (5000 B.C. to 3000 B.C.)," *Facts and Details*.

It was the first culture in the world to domesticate millet.[642] But ca. 5,100 BP a *marked cooling episode* occurred in the Daihai Lake region (China) and precipitated the end of the Yangshao culture.[643]

Figure 174: © Gary Lee Todd
A model of Jiangzhai, a Yangshao village

Uruk, Jemdet Nars, northern Egypt, the northern Sahara, the Knap of Howard, Cucuteni-Trypillia, the Indus civilization and the Yangshao culture all exhibit a similar pattern: after centuries of development, they suddenly collapsed between 5,200 and 4,800 BP.

Most of these collapses are attributed to sudden climate change. However, the collapse of some of these cultures and settlements has been attributed to wars or epidemics.

Notice that climate change, wars and epidemics are not mutually exclusive causes of collapse. *Cometary events can and do cause all three.* Cometary events cause drastic climate changes as detailed above. But they can also be the cause of wars due to the weather-related famine and epidemics that result from cometary-borne pathogens.[644]

About this last point, coincidentally or not, the oldest human carriers of the plague – caused by the *Yersinia pestis* bacteria – were found in Gökhern, Sweden, ca. 4,900 BP.[645] According to the researchers who made the discovery, an epidemic of plague might be the cause of the Neolithic decline, which started around 5,300 BP in Scandinavia:

> It has also been proposed that the **Neolithic decline reached northern European populations in Scandinavia in a process starting around 5,300 BP** (Hinz et al., 2012). Some of the individuals in these studies were found in a passage grave at Frälsegarden in Gökhem parish, Falbygden, western Sweden, which grouped up to 78 individuals **buried between 5100–4900 BP** (based on carbon dating of 34 individuals), **a quite short period of time and large number of bodies compared to other Scandinavian sites** (Ahlström, 2009; Sjögren, 2015). A possi-

---

[642] Houyuan Lua *et al.* (2009), "Earliest domestication of common millet (Panicum miliaceum) in East Asia extended to 10,000 years ago," *Proceedings of the National Academy of Sciences* 106(18):7367–7372.

[643] Jule Xiao *et al.* (2004), "Holocene vegetation variation in the Daihai Lake region of north-central China: a direct indication of the Asian monsoon climatic history." *Quaternary Science Reviews* 23(14–15):1669–1679.

[644] Joseph Rhawn *et al.* (2010), "Comets and Contagion: Evolution, Plague, and Diseases From Space," *Journal of Cosmology* 7:1750–1770.

[645] N. Rascovan *et al.* (2019), "Emergence and Spread of Basal Lineages of *Yersinia pestis* during the Neolithic Decline," *Cell* 176(1–2):295–305.e10.

ble **explanation for the magnitude and short duration of this grave was an epidemic event.**[646]

It has been repeatedly found that microbes are associated with cometary material.[647] Even NASA, through its head of astrobiology research Richard Hoover, has published numerous papers[648] over fifteen years about the presence of bacteria in meteoritic material.

Figure 175: © Hoover / NASA
Fossils of filamentous cyanobacteria found in the Orgueil meteorite

A lot more could be said about cometary material and microbes. The fascinating topic of comets as the *main cause* of mass extinction – removal of obsolete life forms – and the *infusion of viruses*, introducing new and more complex life forms, is addressed in the epilogue.

---

[646]Ibid.

[647]See for example F. Hoyle, and C. Wickramasinghe (2003), "SARS – a clue to its origins?," *The Lancet* 361(9371):1832.

[648]See for example J. Wallis *et al.* (2013), "The Polonnaruwa meteorite: oxygen isotope, crystalline and biological composition," *Journal of Cosmology* 22(2); R. Hoover (2007), "Microfossils of Cyanobacteria in the Orgueil Carbonaceous Meteorite," NASA; Hoover (2008), "Comets, Carbonaceous Meteorites, and the Origin of the Biosphere," in *Biosphere Origin and Evolution*, Springer, pp. 55–68; Hoover (2011), "Fossils of Cyanobacteria in CI1 Carbonaceous Meteorites: Implications to Life on Comets, Europa, and Enceladus," *Journal of Cosmology* 13.

Cometary Venus in Myth

The seven Venusian markers studied above suggest that the first earthly pass of Venus occurred ca. 5,200 BP. Additional spikes in those Venusian markers over the 5,200–4,600 BP period suggest that Venus made other earthly passes during those six centuries.

If those passes coincide with Venus' transformation from a comet to a planet, i.e. a progressive transition from a long-duration highly elliptical cometary orbit to a shorter-duration circular planetary orbit, then we should expect several passes, with shorter and shorter intervals during the 5,200 to 4,600 BP time span.

In this chapter, we will study myths in a search for any hints about cometary Venus in general and its orbital change, number of passes, the effects on Earth of those passes and the time interval between the passes in particular.

The cometary nature of Venus is attested in a number of mythologies, among which are the Aztec and the Maya traditions:

> The Aztec Codex Telleriano-Remensis **represents Venus as a smoking star** in A.D. 1533, linking Venus to imagery of comets (Aveni, 1980). A Maya text in the Songs of Dzitbalche seems to **identify Venus as a smoking star** (Edmonson, 1982).[649]

Sumerian, Egyptian, Semitic, Greek and Hindu traditions also viewed Venus as a comet:

> Each of the goddesses [Inanna, Hathor, Anat, Athena and Kali among others] is explicitly described as a celestial body, **identifiable with the planet Venus; and the imagery surrounding each goddess is consistent with that universally associated with comets** (e.g., long, disheveled hair; serpentine form; identification with a torch; association with eclipses of the Sun, etc).[650]

In addition, in the most archaic texts from Uruk, dated to ca. 5,200 BP, Inanna's name is represented by a pictogram called MUÍ3. A number of specialists consider MUÍ3 as directly inspired by the past cometary aspect of Venus.[651]

Figure 176: © Cochrane
MUI3, the oldest Sumerian pictogram representing Venus

---

[649] Susan Milbrath (1999), *Star Gods of the Maya: Astronomy in Art, Folklore, and Calendars*, University of Texas Press.
[650] Ted Holden (1993), "Velikovsky Update," Efemeral Research Foundation.
[651] E. Cochrane (2011), "Ancient Testimony for a Comet-like Venus," *Proceedings of the NPA*.

Not only was Venus described as a comet, but it was also considered a particularly destructive one, as depicted for example in the prayer of lamentation to Ishtar:

> O Ishtar, queen of all peoples ... Thou art the light of heaven and earth. ... At the thought of thy name the heaven and the **earth quake ...**
> **And the spirits of the earth falter.**
> Mankind payeth homage unto thy mighty name, for thou art great, and thou art exalted.
> All mankind, the whole human race, boweth down before thy power. ...
> How long wilt thou tarry, O lady of heaven and earth...?
> How long wilt thou tarry, **O lady of all fights and of the battle?**
> O thou glorious one, that ... art raised on high, that art firmly established,
> O valiant Ishtar, great in thy might! **Bright torch of heaven and earth, light of all dwellings,**
> **Terrible in the fight, one who cannot be opposed, strong in the battle!**
> **O whirlwind, that roarest against the foe and cuttest off the mighty!**
> **O furious Ishtar, summoner of armies!**[652]

The prayer of lamentations is not an isolated case. Sumerian literature mentions numerous times the destructive nature of Comet Venus (Inanna in the Sumerian tradition):

> [Inanna] was the **blazing, the brilliant fire, I was the blazing fire which became alight in the mountainland; I was the fire whose flame and sparks rained down upon the Rebel land.**[653]

Myths depict Venus as a destructive comet, but do they provide any information about the timing of her passes?

> ... the natives of pre-Columbian Mexico **expected a new catastrophe at the end of every period of fifty-two years and congregated to await the event**. "When the night of this ceremony arrived, all the people were seized with fear and waited in anxiety for what might take place." They were afraid that "it will be the end of the human race and that the darkness of the night may become permanent: the sun may not rise anymore." **They watched for the appearance of the planet Venus, and when, on the feared day, no catastrophe occurred, the people of Maya rejoiced.**
>
> They brought human sacrifices and offered the hearts of prisoners whose chests they opened with knives of flint. On that night, when the fifty-two-year period ended, a great bonfire announced to the fearful crowds that **a new period of grace had been granted and a new Venus cycle started.**
>
> **The period of fifty-two years, regarded by the ancient Mexicans as the interval between two world catastrophes, was definitely related by them to the planet Venus; and this period of Venus was observed by both the Mayas and the Aztecs.**
>
> The old Mexican custom of sacrificing to the Morning Star survived in human sacrifices by the Skidi Pawnee of Nebraska in years **when the Morning Star "appeared especially bright, or in years when there was a comet in the sky"** ...[654]

---

[652] Leonard King (2007), *Enuma Elish: The seven tablets of creation: The Babylonian and Assyrian legends concerning the creation of the world and of mankind*, Cosimo.

[653] Sumerian tablet ASKT 129, lines 11–16. See: James Kinnier-Wilson (1979), *The Rebel Lands*, Cambridge University Press, p. 18.

While the Mayan and Aztec traditions mention a 52-year Venus cycle, we find a similar cycle in *Codex Vaticanus*,[655] one of the oldest copies of the Bible:

> In the *Codex Vaticanus* the world ages are reckoned in multiples of fifty-two years with a **changing number of years** as an addition to these figures. A. Humboldt (*Researches*, II, 28) contraposed the lengths of the world ages in the Vatican manuscript (No. 3738) and their lengths in the system of the tradition preserved by [the Aztec codex] Ixtlilxochitl.[656]

Along the same line there is of course the Judaic 50-year Jubilee tradition as described in Leviticus. The duration of the Jubilee is very close to the Mayan/Aztec tradition and it also refers to the Venus cycle:

> The fiftieth year was a jubilee year ... The festival of the jubilee, with the return of land to its original owners and the release of slaves, bears the character of an atonement, and its proclamation on the Day of Atonement emphasizes this still further. Was there any special reason why fear returned every fifty years? ... **On the Day of Atonement the Israelites used to send a scapegoat to "Azazel"** in the desert. ... **It was also called Azzael, Azza, or Uzza. ... The Arab name of the planet Venus is al-Uzza.**[657]

Other cultures have similar myths about a destructive cycle, but its duration is different. Such is the case in the Etruscan myths. According to Censorinus,[658] it is a 105-year period:

> Four ages of 105 years are referred to by Censorinus (*Liber de die natali*) as having taken place, **according to the belief of the Etruscans**, between world catastrophes presaged by celestial portents.[659]

So, according to Mayan, Aztec and Hebrew traditions, *Venus was a cyclical destructive comet*. Depending on the tradition, the duration of its cycle differs. It is 52 years for the Mayans and the Aztecs, variations of 52 years for *Codex Vaticanus*, 50 years for Leviticus and 105 years in Etruscan tradition.

Besides the cyclical and destructive nature of comet Venus, can we gather more hints about it in ancient mythology? In Mesopotamian mythology, Inanna is Venus, the goddess of war and sex, about whom there is an interesting myth titled *The Descent of Inanna into the Underworld*[660] that goes as follow:

> **Inanna passes through a total of seven gates, at each one removing a piece of clothing or jewelry she had been wearing at the start of her journey, thus stripping her of her power. When she arrives in front of her sister, she is naked:**

---

[654] Velikovsky (1950), pp. 155–156.
[655] Manuscript of the Old and New Testament written in Greek ca. 300 AD. See: Kurt Aland *et al.* (1995), *The Text of the New Testament: An Introduction to the Critical Editions and to the Theory and Practice of Modern Textual Criticism*, William B. Eerdmans Publishing Company, p. 109.
[656] Velikovsky (1950), p. 154.
[657] Leviticus 23:15–16; 25:8–11.
[658] Roman writer from the 3rd century AD. His *De Die Natali*, written in 238 AD, deals with the influence of the stars, religious rites and astronomy.
[659] Velikovsky (1950), p. 153.
[660] Wikipedia Editors (2006), "Inanna," *Wikipedia*.

> "After she had crouched down and had her clothes removed, they were carried away. Then she made her sister Erec-ki-gala rise from her throne, and instead she sat on her throne. Then Anna, the seven judges, rendered their decision against her. They looked at her – it was the look of death. They spoke to her – it was the speech of anger. They shouted at her – it was the shout of heavy guilt. **The afflicted woman was turned into a corpse. And the corpse was hung on a hook.**"

To understand the symbolic meaning of this myth, we have to know that in Mesopotamian mythology and art, the symbolism of nakedness is very specific:

> **Nakedness, correspondingly, is frequently associated with a state of powerlessness and with captivity** and impending execution, not only in Mesopotamian literature but also in art.[661]

If nakedness equates powerlessness and captivity, could the Inanna myth depict the comet Venus *progressively rendered "powerless" and "captured,"* over 7 passes (the 7 gates of the Underworld), into a circular planetary orbit?

This capture of Venus where she is progressively rendered powerless might be reflected by the removal of one her items at each 'gate':

> At the first gate the great **crown** is removed from her head, at the second gate the **earrings** from her ears, at the third gate the **necklace** from her neck, at the fourth gate the **ornaments** from her breast, at the fifth gate the **girdle** from her waist, at the sixth gate the **bracelets** from her hands and feet, and at the seventh gate the covering cloak of her body.[662]

Notice that 6 out of the 7 items listed are jewels, since the girdle removed at the sixth gate is described in the Babylonian version as adorned with birthstones.[663] Of course, the cloak removed at the seventh gate is the symbol of capture/nudity.

Why did Venus, during her descent towards captivity, lose only shiny items? Could this be a symbol for a loss of shiny cometary fragments during each of the six first passes and/or the decrease in overall brightness due to an orbit that is more and more circular?[664]

Coincidentally or not, in the Aztec mythology Venus is Quetzalcoatl,[665] represented as a snake or a dragon – two recurring symbols for cometary bodies.[666]

Often Quetzalcoatl is represented swallowing its own tail like in the picture below. This representation, also known as Ourobouros, symbolizes cycles, particularly of life and death.[667]

Notice that Quetzalcoatl/Ourobouros is usually depicted with seven segments/vertebrae, as indicated by the seven arrows in the picture below:

---

[661] Karen Sonik (2008), "Bad King, False King, True King: Apsû and His Heirs," *Journal of the American Oriental Society* 128(4):737–743.

[662] Manly P. Hall (1928), "The Myth of the Dying God," in *The Secret Teachings of All Ages*, Prabhat Prakashan.

[663] Howard Sherman (2014), *Mythology for Storytellers: Themes and Tales from Around the World*, Routledge, p. 144.

[664] As a general rule, the more elongated the orbit is, the brighter the comet. See: Part III, chapter "Cometary cycle?" and Lescaudron (2014), chapter 18: "Comet Or Asteroid?"

[665] The Editors of Encyclopaedia Britannica (1998), "Quetzalcoatl, Mesoamerican God," *Encyclopaedia Britannica*.

[666] Victor Clube and William Napier (1982), *The Cosmic Serpent: A Catastrophist View of Earth History*, Universe Books, pp. 179–205.

[667] Mircea Eliade (1976), *Occultism, Witchcraft, and Cultural Fashions*, University of Chicago Press, pp. 109–119.

Figure 177: © Aekothewolf
Quetzalcoatl (Venus) and its seven segments

Going back to the Middle East, the Mesopotamians paid very special attention to Venus. It was one of the most venerated deities in the Sumerian pantheon,[668] and the most important and widely venerated in the Assyrian pantheon[669]:

> Ishtar, powerful queen ... is the luminary of heaven and Earth: the greatest Gods have lifted her high, they have made her authority **greatest among the gods** ... they have her heavenly station highest of all whereas at the thought of her name heaven and netherworld quake ... she alone is **"the great one, the exalted one"**.[670]

According to French assyriologist Jean Bottéro, Inanna is the divinity to which the most clay tablets are devoted.[671] Inanna appears in more myths than any other Sumerian deity[672] and it was the most observed astronomical body. One is left to wonder why there was *such an exceptional level of devotion to what mainstream science claims was a rather innocuous planet.*

Depending on the source, Inanna's descent to the underworld is dated between 5,500 BP and 3,900 BP,[673] which fits with the dating of the above-mentioned passes (5,200–4,600 BP). More interestingly, according to professor of comparative mythology Joseph Campbell, Inanna's

---

[668] Gwendolyn Leick (2002), *A Dictionary of Ancient Near Eastern Mythology*, Routledge.
[669] Jeremy Black et al. (1992), *Gods, Demons and Symbols of Ancient Mesopotamia: An Illustrated Dictionary*, British Museum Press.
[670] Jean Bottéro et al. (2001), *Religion in Ancient Mesopotamia*, University of Chicago Press, p. 59.
[671] Ibid.
[672] Charles Penglase (2003), *Greek Myths and Mesopotamia: Parallels and Influence in the Homeric Hymns and Hesiod*, Routledge, chapter 3: "Ninurta."
[673] Wu Mingren (2017), "The Descent of Inanna into the Underworld: A 5,500-Year-Old Literary Masterpiece," *Ancient Origins*.

descent to the underworld was written ca. 4,500 BP, right after the 5,200–4,600 BP destructive time period.[674]

Figure 178: © Creative Commons
Inanna on an Akkadian seal.
She is equipped with 7 spears, a horned helmet and a dress of 7 segments.

Notice that the huge popularity of Inanna described above happened quite suddenly. During the Pre-Sargonic era (ca. 4,300 BP) Inanna had virtually no cult,[675] despite the fact that Inanna was already known for nine centuries:

> **The earliest references to the name Inanna** are on clay tablets from the Eanna district of Uruk; in levels below the remains of major religious buildings dating to the **3rd Dynasty of Ur [ca. 3,200 BC or 5,200 BP]** ...[676]

In summary, if we are to take the above Judaic, Mayan, Aztec, Etruscan and especially Sumerian myths as reflections of actual astronomical events involving Venus, we might expect the following:

- First pass ca. 5,200 BP (first mention of Inanna/Venus)

- Seven passes (the seven rings of the Underworld)

- Decreasing level of destruction and/or brightness (loss of garments and jewels)

- Seventh and last pass ca. 4,600 BP (first mention of Inanna's capture ca. 4,500 BP)

- Time interval between passes is 50 and/or 100 years (Aztec, Mayan, Etruscan and Judaic traditions)

Do geological, geophysical or meteorological data confirm any of those mythical claims? Thanks to the millennial-scale[677] data records studied above, we know that earthly events involving Venus

---

[674] Joseph Campbell (2017), *Creative Mythology*, Joseph Campbell Foundation, chapter 2.
[675] Gwendolyn Leick (2002), *A Dictionary of Ancient Near Eastern Mythology*, Routledge, p. 91.
[676] Collins (1994).
[677] In a millennial scale the basic unit of time is 1,000 years. Most of the graphs in this book use such a large unit of time, which is useful to track long eras but lacks precision to track short (decade-long, for example) events, changes and trends.

happened from 5,200 BP to 4,600 BP, but such a scale doesn't allow a more detailed analysis of what precisely happened *during* those six centuries.

Was it a single event whose effects lasted for six centuries? Was it a series of discrete events? If it's the latter, how many events occurred? At what dates? What was the time interval? What was the magnitude of each event?

Zooming in 5,200–4,600 BP

It's now time to zoom in and to examine high-resolution records. To do so, we have to compile raw data from ice cores.[678]

Below is the bidecadal[679] chart of the average temperature variations from ca. 5,200 BP to 4,600 BP. This average is based on the temperature reconstructions from five regions: the Antarctic, the southern hemisphere, the tropics, the northern hemisphere, and the Arctic.

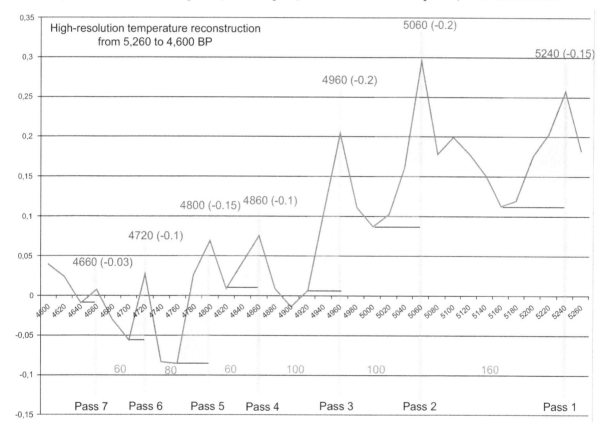

Figure 179: © Sott.net
Ice core temperature reconstruction (5,260 - 4,600 BP)

---

[678] For this purpose we used raw ice core datasets from NOAA and NBI. See: Editors of the WDC (2020), "Ice cores," World Data Center for Paleoclimatology, NOAA, https://www.ncdc.noaa.gov/data-access/paleoclimatology-data/datasets/ice-core; J.P. Steffensen (2020), "Data, icesamples and software," Centre for Ice and Climate, University of Copenhagen, http://www.iceandclimate.nbi.ku.dk/data/.
[679] 20-year increment.

The diagram above reveals that between 5,200 and 4,600 BP, Earth experienced *seven temperature drops* ca. 5,240, 5,060, 4,960, 4,860, 4,800, 4,720 and 4,660 BP as shown by the dates on top of each spike. Those seven temperature drops are consistent with Inanna's descent to the underworld and her passing through the seven gates.

Notice that overall *each pass induces a less severe and less lasting cooling period* as shown by the medium grey triangles to the left of each spike.

For example, pass 1 induced a temperature drop about 15 times more severe and 5 times longer lasting than pass 7. This overall decrease in cooling severity is consistent with *progressively less destructive passes of Venus*, as suggested by the Inanna myth.

Also note the *recurring time intervals* between passes as indicated by the light grey numbers below each trough. We have 60 years between passes 4 and 5, and between pass 6 and 7, which is quite close to the Maya, Aztec and Judaic traditions fixing respectively the recurrence of Venus passes to 52 and 50 years.

On the same note, the 100-year time gap between pass 2 and 3 and between pass 3 and 4 is very close to the 105 years between the passes of Venus in Etruscan mythology.

According to the chart above, four of six time intervals indicate that cometary Venus returned every 60 years and every 100 years. So, maybe the two sets of mythologies – Mesoamerican 52-year cycle, and Etruscan 105-year cycle – were *both right*, only they were referring to different passes of cometary Venus.

Also the interval between each of the seven passes tends to diminish overall from 160 years between the first and second passes to 60 years between the sixth and final passes. This overall interval decrease is consistent with a *progressive capture* of cometary Venus in the solar system, where its orbit progressively becomes shorter and more circular.

While the time gap between each pass decreases overall, there is however one exception: the time gap between passes 5 and 6 is longer (80 years) than the previous time gap between passes 4 and 5 (60 years).

This non-linear decrease in time intervals between the passes of cometary Venus might be due to the fact that comets, even short-period/stable ones, do not return at exact periods because of perturbations caused by astronomical bodies, particularly large planets like Jupiter.[680]

This variability even applies to the most famous comet of our era, Halley's Comet, which has an *average* period of nearly 77 years,[681] but whose single periods span from 74.33 years to 79 years.[682]

Researcher Joachim Seifert came up with a temperature graph similar to the one above, but he added one variable,

Figure 180: © European Southern Observatory
Halley's Comet photographed during its 1986 pass. Its next pass is expected in 2061.

---

[680] David Hughes (2003), "The variation of short-period comet size and decay rate with perihelion distance," *Monthly Notices of the Royal Astronomical Society* 346(2):584–592.
[681] F.W. Henkel (1908), "Halley's Comet," *Popular Astronomy* 15:238–245.
[682] H.C. Wilson (1908), "The Next Apparition of Halley's Comet," *Popular Astronomy* 16:265–270.

the Earth Orbital Oscillation (EOO),[683] i.e. the temperature changes induced by the variation in Earth's orbit and the subsequent changes and Earth–Sun distance.

Because of the limited temperature variations induced by this EOO variable, it is negligible when dealing with major events on a millennial scale, but it can relevant when dealing with high-resolution temperature reconstructions:

> ... the upper and the lower [horizontal lines in the graph are the] **Earth orbital oscillation line, within which the Earth climate varies, if not impacted by large cosmic bolides.** As we demonstrate, the Holocene temperature evolution does not remain confined within these upper and lower horizontal lines, because **strong cosmic impacts always and necessarily produce a strong temperature down-spin spike**, followed by a strong upward temperature rebound spike, regressing thereafter. This is the so-called Z-shaped temperature pattern of each cosmic impact on Earth.[684]

Here is Seifert's temperature graph:

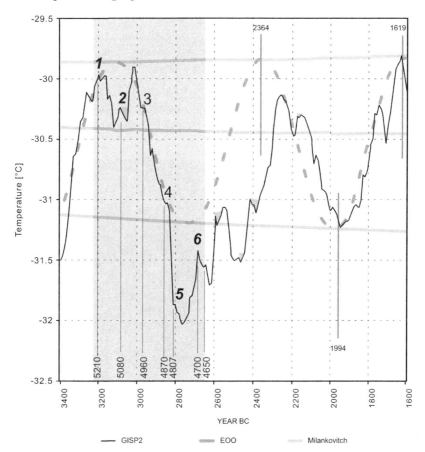

Figure 181: © Seifert
EOO temperature reconstruction (3,400–1,600 BC)

---

[683]Sean Raymond (2014), "Real-life sci-fi worlds #3: the oscillating Earth," *Planetplanet*.
[684]Seifert and Lemke (2016b).

In the graph above, the time period we're studying – ca. 5,200–4,600 BP – is in the medium grey rectangle. The medium grey dotted line is the theoretical Earth temperature if the only driver was the Earth Orbit Oscillation (EOO), hence its sinusoidal shape. But we can see that the recorded temperature as shown by the solid black curve *departs from the EOO curve in several places*.

Seifert lists 4 catastrophic cooling events that caused four of those departures, namely events 1, 2 5 and 6 as indicated by the bold italics figures in the diagram above.

- **Event 1 ca. 5,210 BP (3,210 BC)**. According to Seifert, this event is related to the Andaman Gulf impact:

> The BC 3200 event is recognized in lake filling data, both at Lake Accesa and Lake Constance, as described before. **The BC 3200 cosmic impact produced a Z-shaped temperature pattern, which lasts until 2900 BC**. This cosmic impact is responsible for the delayed temperature peak at 3000 BC, displacing the regular peak at 3081 BC by 80 years.
>
> Looking for prospective impact candidates of this time, we found a **cosmic meteor impact, striking the Andaman Sea**. At Cape Pakarang (west coast of Thailand), a **mega-tsunami** struck (Neubauer, 2011) at 3200 BC, as an outstanding megatsunami event. Regular seaquake tsunamis are not forceful enough to destroy reefs and to move enormous cut-off boulders far inland.[685]

The Andaman impact is not the only suspect to explain this major cooling event. According to engineer Alan Bond,[686] ca. 5,123 BP a cometary impact caused the largest Alpine landslide in crystalline bedrock, known as the Köfels[687] event, which moved an enormous volume of material exceeding 3.2 km^3.[688]

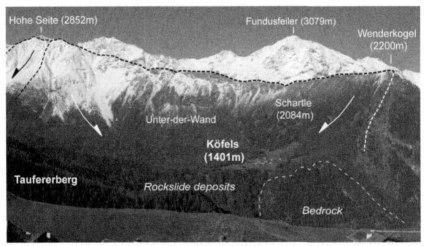

Figure 182: © Prager
Morphology of the Köfels rockslide

---

[685] Joachim Seifert and Frank Lemke (2016b), "Climate Pattern Recognition In The Mid-To-Late Holocene (4800 BC To 2800 BC, Part 3)," *ResearchGate*.
[686] Alan Bond et al. (2008), *A Sumerian Observation of the Köfels' Impact Event*: A Monograph, Alcuin Academics.
[687] Ötz valley, Tyrol, Austria.
[688] Christoph Prager et al. (2009), "Geological controls on slope deformations in the köfels rockslide area (Tyrol, Austria)," *Austrian Journal of Earth Sciences* 102:4–19.

- **Event 2 ca. 5,080 BP (3,080 BC).** Seifert identified a second departure from the EOO curve that he attributes this time to a potential mega-eruption, as indicated by the dark grey trough in the diagram above.

Seifert attributes to the ca. 5,080 BP event the greatest sulfur atmospheric release of the past 10,000 years,[689] which we described earlier in this book.[690] Notice that, as indicated in the table below, this alleged 'mega-eruption' has not yet been convincingly attributed to any known volcanic eruption:

Camp Century		BC			
Unknown	?	50 ± 30	192	1?	(120)
Unknown	?	210 ± 30	72	1?	(45)
Unknown	?	260 ± 30	54	1?	(35)
Hekla 3	Iceland	1120 ± 50	99	1	60
Thera	Greece	1390 ± 50	98	2	125
Hekla 4	Iceland	2690 ± 80	96	1	60
Unknown	?	3150 ± 90	255	1?	(160)
Mt Mazama	Oregon, USA	4400 ± 110	156	2	200
Hekla 5/Thjórsá flow	Iceland	5470 ± 130	90	1	60
Unknown	?	6060 ± 140	119	1?	(75)
—	?	6230 ± 140	102	1?	(65)
—	?	7090 ± 160	79	1?	(50)
—	?	7240 ± 160	124	1?	(80)
—	?	7500 ± 160	51	1?	(35)

Figure 183: © Hammer
Major sulfur spikes over the past 9,500 years

In addition to this suspected "mega-eruption," a series of cosmic impacts were documented about this time. The luminescence and carbon dating provide the same dating ca. 3,000 BC (5,000 BP) for the Morasko Crater in Poland and the Kaali, Ilumetsa and Tsõõrikmäe craters in Estonia.[691]

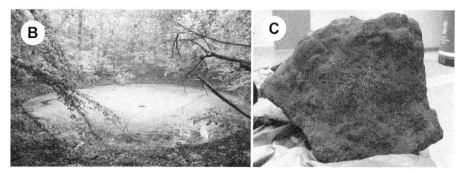

Figure 184: © Zwolinski
B – One of the Morasko impact craters
C – The largest fragment of Morasko meteorite weighing 261 kg.

---

[689] 255 kg/km² of sulfuric acid (H_2SO_4) fallout, recorded in the ice cores of Camp Century, Greenland.
[690] See chapter titled "Sulfur Dioxide."
[691] Wojciech Stankowski *et al.* (2007), "Luminescence Dating of the Morasko (Poland), Kaali, Ilumetsa and Tsõõrikmäe (Estonia) Meteorite Craters," *Geochronometria* 28(1):25–29.

The impact field in Morasko contains eight crater impacts. Large meteorites were found in some of these craters[692] and a peat bog nearby revealed meteorite metallic spherules.[693]

- **Event 5 ca. 4,807 BP (2,807 BC).** This event is associated by Seifert to the Burckle impact:

> The date 2807 BC is given in Chinese celestial observation records. This **impact was enormous in size and effect, the impact crater is 20 km in diameter**. This impact sent global temperatures instantly deep down, the enormous fall-out of atmospheric moisture produced widespread global inundations.[694]

4,807 BP is also the suggested date for an asteroid or comet impact occurring between South Africa and Antarctica,[695] possibly causing the Burckle crater and the Fenambosy Chevron.[696]

Figure 185: © Weaver
Fenambosy chevron, Madagascar. Black arrows depict the front edge of the 150-meter (490-foot) high plateau escarpment that was overtopped by the megatsunami wave.

---

[692]Zbigniew Zwolinski *et al.* (2017), "Existing and Proposed Urban Geosites Values Resulting from Geodiversity of Poznań City," *Quaestiones Geographicae* 36(3):125–149.
[693]Wojciech Stankowski (2011), "Luminescence and radiocarbon dating as tools for the recognition of extraterrestrial impacts," *Geochronometria* 38(1):50–54.
[694]Seifert (2016b).
[695]Damien AtHope (2018), "Explaining My thoughts on the Evolution of Religion," *Damienmarieathope.com*
[696]For more details about the Fenambosy Chevron, see: W. Masse, R. Weaver, *et al.* (2007), "Missing in Action? Evaluating the Putative Absence of Impacts by Large Asteroids and Comets during the Quaternary Period," *ResearchGate*.

- **Event 6 ca. 4,700 BP (2,700 BC)**. This event is associated by Seifert with the Campo del Cielo[697] impact:

> The literature (Barrientos, 2014) actually sets a time frame of 2840–2146 BC, but the only impact date remaining is at 2700 BC. **This impact is small to medium, delaying the temperature recovery after the Burckle event by one century.**[698]

In addition to the four departures from the EOO curve spotted by Seifert and described above, there are three additional sudden cooling events:

- **Event 3 ca. 4,960 BP**, as indicated by the number 4 in the EOO diagram,
- **Event 4 ca. 4,870 BP**, as indicated by the number 3 in the EOO diagram,
- **Event 7 ca. 4,650 BP**, as indicated by the number 7 in the EOO diagram.

Figure 186: © Livio Gutierrez
The 30-ton meteorite found in Campo del Cielo

The Seifert EOO diagram is based on Greenland ice cores,[699] while our duodecenal diagram is based on the average temperature reconstruction from five regions: the Antarctic, the southern hemisphere, the tropics, the northern hemisphere, and the Arctic. Despite the use of different sources, both diagrams provide a strikingly similar image – *seven temperature drops with virtually the same timing* – as shown by the following table:

		Event 1	Event 2	Event 3	Event 4	Event 5	Event 6	Event 7	
Seifert	Greenland (GISP2)	5210	5080	4960	4870	4807	4700	4650	
Duodecanal	5 regions average	5240	5060	4960	4860	4800	4720	4660	
Difference (years)		-30	20	0	10	7	-20	-10	13,80

Figure 187: © Sott.net
GISP2 vs. regional average temperature reconstruction

For the seven events described previously, the dating difference between the GISP2 temperatures and the regional average reconstruction is *only 13.8 years*. It is not a bad match at all for events that happened about 5,000 years ago, knowing that the error margin for ice core dating can reach as much as 2%,[700] i.e. about 100 years, for events that occurred in that time period.

---

[697] Campo del Cielo, Argentina, refers to a crater field covering almost 60 km² (23 sq. mi.) and containing at least 26 craters. The largest two meteorites fragments are the 30.8-ton Gancedo and 28.8-ton El Chaco, which are some of the heaviest meteorites ever found on Earth.
[698] Seifert (2016b).
[699] GISP2, Greenland Ice Sheet Project 2.
[700] G.A. Zielinski (1994), "Record of Volcanism Since 7000 B.C. from the GISP2 Greenland Ice Core and Implications for the Volcano-Climate System," *Science New Series* 264(5161).

Conclusion

Most literature dealing with cometary events posits regular cycles or one-time events. While these are often true, they're not the whole picture. The seven passes of Venus described above were neither a one-time event nor part of a constant cycle.

Cometary events can be ongoing or a thing of past. Likewise they can be periodic, pseudo-periodic or a one-shot event.

For example, we know of ongoing periodic cometary cycles like the 27.9-million-year cycle followed by Nemesis[701] and its accompanying cometary swarm, or the 3,600-year cometary cycle described previously.

There are also ongoing pseudo-periodic cycles like Comet Halley, whose average period is 77 years, but whose single periods span from 74.33 to 79 years.

There are seemingly one-shot events like the 12,900 BP cometary event described earlier in this book.[702]

And finally there are past pseudo-periodic comets like cometary Venus from 5,200 BP to 4,600 BP with a decreasing orbital period: from 160 years for pass 1 to 60 years for pass 7.

In each of the four parts of this book, we dealt with past cometary events. All of these refer to ancient history; cometary events seem so remote when observed from a human timescale.

However, in 2013 the Chelyabinsk event reminded us that space rocks are very real.

The Chelyabinsk overhead explosion released 40 times more energy than the Hiroshima bomb and was 30 times brighter than the Sun. It spread destruction over 500 km² (190 square miles) and damaged more than 7,000 buildings.[703] It was only eight years ago and, despite the magnitude of the event, *nobody saw it coming*.[704]

Figure 188: © AP Photo/Nasha gazeta
Dashboard camera video grab of the Chelyabinsk meteor shot on Feb. 15, 2013

---

[701] Lescaudron (2014), chapters 13 to 19.

[702] Part 1, "Of Flash Frozen Mammoths and Cosmic Catastrophes."

[703] Prabhjote Gill (2019), "The last major asteroid to hit Earth destroyed 500 square kilometers – and it could have been much worse," *Business Insider*.

[704] Ruchika Agarwal *et al.* (2016), "The unsettling reasons no one saw the Chelyabinsk meteor over Russia coming – and why it could happen again," *Business Insider*.

Even more recently, in March 2020, a meteorite impacted Akure, Nigeria, creating an 8-meter deep,[705] 21-meter wide[706] crater and destroying 70 buildings.[707]

Figure 189: © PMNews Nigeria
The crater left by a meteor impact in Akure, Nigeria

The Chelyabinsk and Akure events remind us that cometary events are not just an abstract concept that belongs exclusively to the distant past.

---

[705] 26 feet.
[706] 69 feet.
[707] PM News Nigeria Editors (2020), "Akeredolu lied, Akure blast caused by meteorite: Professor Adepelumi," *PM News Nigeria*.

Epilogue

In *Earth Changes and the Human–Cosmic Connection*, we saw that life on Earth is punctuated by mass extinctions, most of which were due to cyclical cometary bombardments,[708] as illustrated in the graph below, which shows that out of 19 events (black circled dots), 11 (grey circled dot) lie on the vertical lines showing 27-million-year intervals:

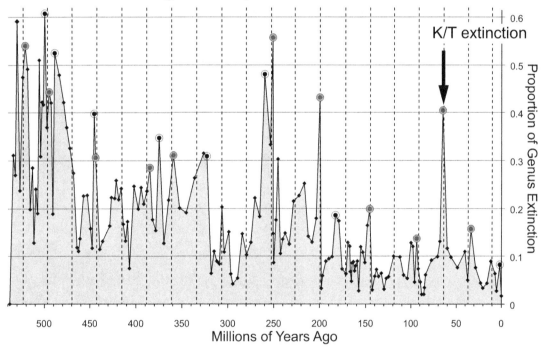

Figure 190: © Sott.net adapted from Melott & Bambach
Plot of extinction intensity and cyclicity from the Sepkoski data.

For years, I considered such cometary encounters solely as agents of mass destruction. But while studying the Tunguska event as part of the research for this book, I stumbled upon the following:

> It was found that the **genotypic dispersion has sharply increased in the Tunguska trees.** The effect is prominent, has a patchy character and **concentrates toward the epicenter area, as well as toward of the TSB [Tunguska Space Body] trajectory projection** (Vasilyev 1999, 2000, 1998). **At maximums the genotypic dispersion shows about 12-fold increment** (Vasilyev 2000). One of the maximums coincides again with the Mount Chirvinskii, another – with the calculated center of the light flash.[709]

---
[708] Lescaudron (2014), chapter 15 : "Enter Nemesis."

While increased mutations are compatible with cometary encounters being exclusively agents of destruction, the Tunguska trees revealed another peculiarity:

> The cause of the **anomalous growth of tree rings after 1909** is more controversial. We collected tree ring data for 9 spruces, 1 larch and 1 Siberian pine. A comparison of the average tree ring width over about 30 years before 1907, and exactly the same period after 1909, has confirmed the width increase for all the 11 trees examined.
>
> From these data no correlation with the tree position has been found. The trees were divided into two groups: 5 trees with an average ring width before 1907 equal to about 0.4 mm and a second group with a ring width of about 1 mm. **After 1909 both groups reach approximately the same ring width of about 1.2-1.5 mm with an increase for the first group by a factor 3-4, as against a factor 1.2-1.5 for the second group.**[710]

Increased growth doesn't sound like random detrimental mutation but rather *beneficial* mutation. Could cometary mass extinction improve life forms? Maybe Tunguska was an isolated case, so I had a look at other documented cometary-induced mass extinctions, in particular the well-documented K/T boundary,[711] when 66 million years ago, the Chicxulub impact wiped out the dinosaurs along with 75% of all species on the planet,[712] as shown by the black arrow in the cyclical cometary extinction diagram above.

The K/T boundary is a geological term referring to a very thin strata of clay[713] – shown by the white arrow in the picture to the right – sandwiched between the sediments of the Tertiary above and the sediments of the Cretaceous below.

Figure 191: © Creative Commons
The K/T boundary

Right below the K/T boundary lie the fossils of the dinosaurs and right above it – meaning right after the mass extinction – lie the fossils of *fully developed new species* with no known ancestors:

> Immediately following the end-Cretaceous, primitive birds enjoyed a **surge of development**, and large flightless birds like today's ostrich suddenly appeared. Tree sloths, armadillos and anteaters

---

[709] N.V. Vasilyev (1999), "Ecological consequences of the Tunguska catastrophe," *in Problemi radioekologii i pogranichnikh discipline* 89.

[710] See: G. Longo *et al.* (1994), "Search for microremnants of the Tunguska cosmic body," *Planetary and Space Science* 42(2):163–177; R. Serra *et al.* (1994), "Experimental hints on the fragmentation of the Tunguska cosmic body," *Planetary and Space Science* 42(9):777–783.

[711] Cretaceous/Tertiary.

[712] K. Kaiho *et al.* (2016), "Global climate change driven by soot at the K-Pg boundary as the cause of the mass extinction," *Scientific Reports* 6(28427).

[713] About 1.2 cm (half an inch).

saw dramatic development, as did egg-laying mammals such as the platypus and the echidnor. Other mammals, until then not much bigger than a house cat, took off in a **rapid and spectacular diversification, all with no known ancestors.**

Sea-dwellers enjoyed the same fate. Twenty-one of the twenty-seven species of lampshells (brachiopods) were completely obliterated at the K-T boundary, only to be **suddenly replaced by twenty-four entirely new species.**[714]

The K/T boundary is not an oddity. *Virtually every geological period exhibits the same pattern.* It starts with a mass extinction quickly followed by new, fully developed, life forms that appear 'out of thin air':

> Look at the Ordovician, when **jawless fish with no known ancestors suddenly appeared**; or the Silurian, **when algae crawled out of the sea** and onto the barren ground; or the Devonian, when **coniferous trees suddenly appeared, as did ferns, seemingly out of thin air** ... So many new kinds of fish appeared in the Devonian that it's called the "Age of the Fishes". **Sharks appeared suddenly, and the first amphibian, Ichthyostega, crawled out of the water** and onto the land.[715]

The correlation between the aftermath of mass extinction and new life forms is stunning. In fact almost all major living branches of life appeared right after a mass extinction (see figure on next page).

The above clearly suggests that *major cometary impacts are not only acts of destruction through the removal of obsolete life forms during mass extinctions, but also creative acts through the introduction of more elaborate life forms.*

So the question becomes: How can cometary impacts introduce new and more complex life forms? This question lingered in my mind until the COVID hysteria, about which I wrote an article.[716] During this process I learnt a lot about viruses, including their pervasive presence in the genetic code of all life forms, including humans:

> One of the most earth-shaking papers of this century was the publication of the human genome sequence. **About half, possibly even two-thirds of the sequence are composed of more or less complete endogenous retroviruses (ERVs) and related retroelements (REs)** ... The origin of REs is being discussed as remnants of ancient retroviral germline infections that became evolutionarily fixed in the genome.
>
> About 450,000 human ERV (HERV) elements constitute about 8% of the human genome consisting of hallmark retroviral elements like the gag, pol, env genes and flanking long terminal repeats (LTR) that act as promoters. Howard Temin, one of the discoverers of the reverse transcriptase, in 1985 already described endogenous retrovirus-like elements, which he estimated to about 10% of the human and mouse genome sequence.
>
> **The actual number is about 45% as estimated today. In some genes, such as the Protein Kinase Inhibitor B (PKIB) gene, we determined about 70% retrovirus-related sequences. Is there a limit? Could it have been 100%?**[717]

---

[714] Robert Felix (2008), *Magnetic Reversals and Evolutionary Leaps: The True Origin of Species*, Sugarhouse Publications, p. 33.
[715] Ibid.
[716] Pierre Lescaudron (2020), "Compelling evidence that SARS-CoV-2 was manmade," *Sott.net*.

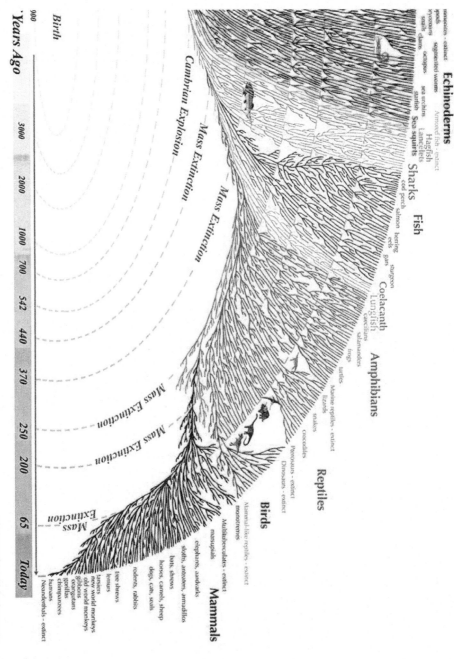

Figure 192: © Eisenberg
Mass extinctions and new forms of life.

Not only does viral DNA constitute a large part of our genetic code, it has been demonstrated that this 'foreign' DNA offers new beneficial functions to the host:

> In 1996, Roy J Britten, of the California Institute of Technology, was able to list **ten examples in which endogenous retroviral sequences helped regulate the expression of a useful gene**. Seven of the ten examples were human.[718]

Actually, the role played by viral DNA is so critical that hosts' reproductions would be simply impossible without it:

> ... when Corrado Spadafora, the Italian researcher who has produced such pioneering work in cancer research, applied an anti-reverse transcriptase drug in this very early stage of embryogenesis in the mouse, **all development ceased at the four-cell stage. It would appear that endogenous retroviruses are playing a very profound, if completely unknown, role at this early stage in mammalian embryogenesis.**[719]

In the light of the quote above, viruses seem more fundamental than even life itself; *they are the information carriers – genetic codes – from which biological life stems*. Viruses are more than life: they are the very source of life; they are the informational precursors, the initiators of life forms.

Knowing the central role played by viruses in the genetic code of life forms, could it be that it is the infusion of new viruses that triggers the evolutionary leaps that mark the aftermath of cometary-induced mass extinctions?

The fact is that cometary material is unexpectedly rich in organic material. Fifty-two amino acids have been found in Murchison meteorites,[720] while there are only 19 of them naturally occurring on our planet. More than that, as already noted, several researchers including NASA's head of astrobiology research Richard Hoover,[721] have published a number of papers about the presence of bacteria in meteoritic material (see image on next page).

Not only do meteorites carry microbes, the Earth's upper atmosphere is also laden with microorganisms. Dozens of types of microorganisms, including three species[722] – two bacterial and one fungal – that are unknown on Earth have been found as high as 41 km,[723] where no air from lower down would normally be transported.[724]

From the above, we are left with two non-mutually exclusive hypotheses to explain the infusion of new viruses on Earth: meteorites directly carrying new viruses, and meteorites puncturing the atmosphere and enabling the entrance of new viruses.

---

[717] K. Moelling and F. Broecker (2019), "Viruses and Evolution – Viruses First? A Personal Perspective," *Frontiers in microbiology* 10:523.

[718] Frank Ryan (2013), *Virolution*, Harper Collins UK.

[719] Ibid.

[720] J.R. Cronin *et al.* (1983), "Amino acids in meteorites," *Advances in Space Research* 3(9):5–18.

[721] See chapter "Impact on Human Populations."

[722] S. Shivaji *et al.* (2009), "Janibacter hoylei sp. nov., Bacillus isronensis sp. nov. and Bacillus aryabhattai sp. nov., isolated from cryotubes used for collecting air from upper atmosphere," *International Journal of Systematic and Evolutionary Microbiology* 59(Pt 12):2977–86.

[723] 25 miles.

[724] J.V. Narlikar *et al.* (2003), "Balloon experiment to detect micro-organisms in the outer space," *Astrophysics and Space Science* 285(2):555–62.

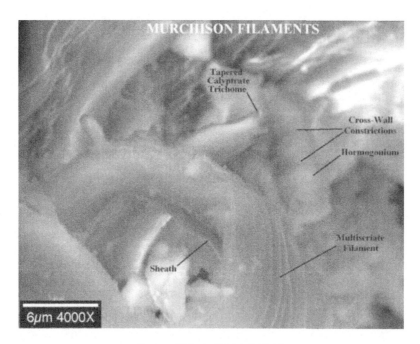

Figure 193: © NASA/MSFC
Cyanobacterial filaments in the Murchison CM2 meteorite

At this point one last question remains: how does viral DNA shape new, complex life forms? Mainstream science postulates that DNA is the code that controls the synthesis of proteins – the building blocks of life. Different DNA = different code = different proteins = different morphology. It's a simple and neat explanation, but is it true?

How do we explain that, in one host, every single cell carries the same DNA, but those cells follow different developmental paths? One becomes muscle cell, another one a neuron, a third one a bone cell. How to explain that when a salamander loses a limb, it grows a perfect replica of it? *Where is the template, where is the blueprint of the whole limb?*[725]

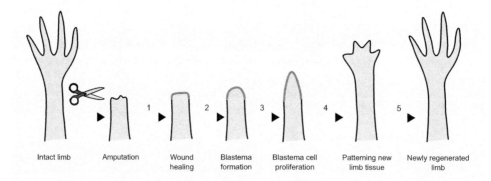

Figure 194: © Whited
Steps in salamander limb regeneration

---

[725]See for example: Rupert Sheldrake (2009), *Morphic Resonance: The Nature of Formative Causation*, Inner Traditions/Bear.

I dedicated the last part of *Earth Changes and the Human–Cosmic Connection* to demonstrating the existence of a non-local information field, including morphology-related information.[726] I won't transcribe here all 100 pages, but one excerpt might suffice to convey the main idea:

> There's a young student at this university who has an **IQ of 126, has gained a first-class honors degree in mathematics, and is socially completely normal. And yet the boy has virtually no brain** ... When we did a brain scan on him, we saw that instead of the normal 4.5-centimeter thickness of brain tissue between the ventricles and the cortical surface, there was just a thin layer of mantle measuring a millimeter or so. **His cranium is filled mainly with cerebrospinal fluid.**[727]

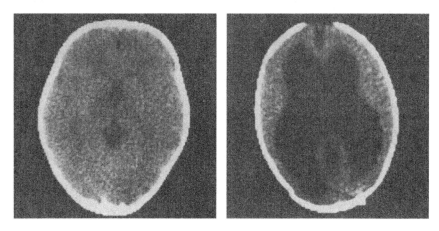

Figure 195: © Lorber
(Left) A horizontal scan shows the brain ventricles in a normal individual
(Right) Large cavities in a hydrocephalic patient

An individual virtually devoid of any brain and yet fully functional intellectually and socially suggests that our ideas, thoughts, memories might be *non-local* instead of being stored/created in our brain as usually assumed.

That's the main premise of the concept of an 'information field' that contains all information, including that relating to the morphology of life forms.

So if information, including morphic information, is non-local and the acquisition of new viral DNA enables morphologically more complex life forms, then *viral DNA must be able somehow to modulate the connection of life forms to the information field.*

Could viruses act like antennae? Viruses are DNA and DNA was recently revealed to have all the features of a fractal antenna when exposed to electromagnetic fields:

> The wide frequency range of interaction with EMF is the functional characteristic of a fractal antenna, and **DNA appears to possess the two structural characteristics of fractal antennas, electronic conduction and self-symmetry**. These properties contribute to greater reactivity of DNA with EMF in the environment ...[728]

---

[726] "Part IV: Role of the elites, the human-cosmic connection," pp. 192–290.

[727] Roger Lewin (1980), "Is your brain really necessary?," *Science* 210(4475):1232–4.

[728] M. Blank and R. Goodman (2011), "DNA is a fractal antenna in electromagnetic fields," *International Journal of Radiation Biology* 87(4):409–15.

So, do new viruses enable in their host an enhanced connection to the information field? Are major cometary events the window of opportunity that 'intelligent design' uses to remove obsolete life forms through mass extinction and to introduce more elaborate life forms via the new DNA carried by viruses?

The answer to those two questions and the general content of this epilogue might very well be the topic of a next book.

Bibliography

Meteoritical Society – International Society for Meteorites and Planetary Science. "Meteoritical Bulletin Database." https://www.lpi.usra.edu/meteor/.

National Geographic Resource Library. "Continental shelf." *National Geographic Encyclopedic Entry.* https://www.nationalgeographic.org/encyclopedia/continental-shelf/.

Stoneridge Engineering. "What are Lichtenberg figures, and how do we make them?" http://www.capturedlightning.com/frames/lichtenbergs.html.

Velikovsky Encyclopedia Editors. "Electric Sun Model." *The Velikovsky Encyclopedia.* https://www.velikovsky.info/electric-sun-model/.

Velikovsky Encyclopedia Editors. "Worlds in Collision." *The Velikovsky Encyclopedia.* https://www.velikovsky.info/worlds-in-collision/.

Agarwal, Ruchika *et al.* (2016). "The unsettling reasons no one saw the Chelyabinsk meteor over Russia coming – and why it could happen again." *Business Insider.*

Aland, Kurt *et al.* (1995). *The Text of the New Testament: An Introduction to the Critical Editions and to the Theory and Practice of Modern Textual Criticism.* William B. Eerdmans Publishing Company.

Allan, D., and Delair, J. (1997). *Cataclysm!.* Bear & Co.

Allen, Harris (2011). *Somewhere in the Bible: Understanding Bible Scriptures and Creation.* WestBow Press.

Alvarez, L. (1980). "Extraterrestrial Cause for the Cretaceous-Tertiary Extinction." *Science* 208(4448):1095–1108.

American Museum of Natural History (2016). "Where is Earth's Water?" *Amonh.org.*

American Physical Society's Division of Fluid Dynamics (2015). "How does fur keep animals warm in cold water?" *ScienceDaily.*

AMS Editors (2020). "Events stats." American Meteor Society Website. https://fireball.amsmeteors.org/members/imo_fireball_stats/.

Andrea, Alfred J. (2011). *World History Encyclopedia, Volume 10.* ABC-CLIO.

Anthony, David W. (2010). *The Horse, the Wheel, and Language: How Bronze-Age Riders from the Eurasian Steppes Shaped the Modern World.* Princeton University Press.

Armstrong, Michael (2008). "The Grand Canyon: Part One." *Thunderbolts.info.*

Armstrong, Terence (1971). "Bellingshausen and the discovery of Antarctica." *Polar Record* 15(99):887–889.

AtHope, Damien (2018). "Explaining My thoughts on the Evolution of Religion." *Damienmarieathope.com.*

Avery, Appiah (2008). "Mammuthus primigenius – The Woolly Mammoth." *McMaster University.*

Bagley, Katherine (2009). "Scientists Find Smoking Gun for California Comet Theory." *National Audubon Society.*

Baillie, Mike *et al.* (1988). "Irish tree rings, Santorini and volcanic dust veils." *Nature* 332:344–346.

Baldini, J.U.L. *et al.* (2018). "Evaluating the link between the sulfur-rich Laacher volcanic eruption and the Younger Dryas climate anomaly." *Climate of the Past* 14(7):969–990.

Barbiero, Flavio (2020). "Changes in the Rotation Axis of Earth After Asteroid/Cometary Impacts and their Geological Effects." *International Journal of Anthropology* 35(1-2).

Bargmann, V. *et al.* (1962). "On the Recent Discoveries Concerning Jupiter and Venus." *Science* 138(3547):1350–1352.

Bast, Robert (2009). "Crustal Poleshifts." *Survive2012*.

Beals, Carlyle (1955). "Earth Impact Database." *Dominion Observatory*.

Becker, L. *et al.* (2009). "Wildfires, Soot and Fullerenes in the 12,900 ka Younger Dryas boundary layer in North America." *American Geophysical Union*.

Becker, Luann *et al.* (2001). "Impact Event at the Permian-Triassic Boundary: Evidence from Extraterrestrial Noble Gases in Fullerenes." *Science* 291(1530).

Berkeley Editors (2020). "About Mammoths." *University of California in Berkeley*.

Bertaux, J. *et al.* (1989). "Deuterium content of the Venus atmosphere." *Nature* 338:567–568.

Black, Jeremy *et al.* (1992). *Gods, Demons and Symbols of Ancient Mesopotamia: An Illustrated Dictionary*. British Museum Press.

Blagrave, K.P.M.A. *et al.* (2006). "Photoionized Herbig-Haro Object in the Orion Nebula." *The Astrophysical Journal*.

Blank, M., and Goodman, R. (2011). "DNA is a fractal antenna in electromagnetic fields." *International Journal of Radiation Biology* 87(4):409–15.

Bond, Alan *et al.* (2008). *A Sumerian Observation of the Köfels' Impact Event*: A Monograph. Alcuin Academics.

Bos, J.A. *et al.* (2017). "Multiple oscillations during the Lateglacial as recorded in a multi-proxy, high-resolution record of the Moervaart palaeolake (NW Belgium)." *Quaternary Science Reviews* 162:26–41.

Bottéro, Jean *et al.* (2001). *Religion in Ancient Mesopotamia*. University of Chicago Press.

Boulton, G.S. *et al.* (2004). "Evidence of European ice sheet fluctuation during the last glacial cycle." *Developments in Quaternary Sciences* 2(1):441–460.

Bradford, Alina (2016). "Facts About Woolly Mammoths." *Live Science*.

Bradley, Raymond (1988). "The explosive volcanic eruption signal in northern hemisphere continental temperature records." *Climatic Change* 12(3):221–243.

Brami, Maxime (2014). "Revisiting Hacılar." *Arkeoloji ve Sanat Dergisi* 146(13).

Brook, Ed *et al.* (2015). "Isotopic constraints on greenhouse gas variability during the last deglaciation from blue ice Archives." in Michael Sarnthein *et al.*, *Deglacial changes in ocean dynamics and atmospheric CO2: modern, glacial, and deglacial carbon transfer between ocean, atmosphere and land*, Leopoldina Symposium, Nova Acta Leopoldina, pp. 39–43.

Brooks, Nick (2012). "Beyond collapse: Climate change and causality during the Middle Holocene Climatic Transition, 6400–5000 years before present." *Geografisk Tidsskrift-Danish Journal of Geography* 112:93–104.

Bruce, Charles (n.d.). "Comets." *Plasma-Universe.com*.

Buis, Alan (2019). "The Atmosphere: Getting a Handle on Carbon Dioxide." NASA Global Climate Change.

Buizert, C. *et al.* (2014). "Greenland temperature response to climate forcing during the last deglaciation." *Science* 345 (6201): 1177–1180.

Bunch, Ted *et al.* (2012). "Melted glass from a cosmic impact 12,900 years ago." *Proceedings of the National Academy of Sciences* 109(28):E1903-E1912.

Burke, John G. (1986). *Cosmic Debris: Meteorites in History*. University of California Press.

Burns, R. *et al.* (2018). "Direct isotopic evidence of biogenic methane production and efflux from beneath a temperate glacier." *Scientific Reports* 8(17118).

Burr, Devon *et al.* (2009). *Megaflooding on Earth and Mars*. Cambridge University Press.

Busemann, Henner *et al.* (2009). "Ultra-primitive interplanetary dust particles from the comet 26P/Grigg–Skjellerup dust stream collection." *Earth and Planetary Science Letters* 288(1–2):44–57.

Butrica, Andrew J. (1996). "To See the Unseen: A History of Planetary Radar Astronomy." *NASA SP-4218*.

Cai, C. *et al.* (2018). "Water input into the Mariana subduction zone estimated from ocean-bottom seismic data." *Nature* 563:389–392.

Cain, Fraser (2008). "Craters on Venus." *Universe Today*.
Cain, Fraser (2013). "What is the Hottest Planet in the Solar System?" *Universe Today*.
Cain, Fraser (2016). "What is the Earth's Mantle Made Of?" *Universe Today*.
Campbell, Bruce et al. (2019). "The mean rotation rate of Venus from 29 years of Earth-based radar observations." *Icarus* 332:19–23.
Campbell, Joseph (2017). *Creative Mythology*. Joseph Campbell Foundation.
Campbell, Stuart (2007). "Rethinking Halaf Chronologies." *Paléorient* 33(1):103–136.
Canright, Shelley (2009). "Escape Velocity: Fun and Games." *NASA website*.
Carr, Michael (2006). *The Surface of Mars*. Cambridge University Press.
Carrier, Richard (2014). *On the Historicity of Jesus: Why We Might Have Reason for Doubt*. Sheffield Phoenix Press.
Casati, Michele (2014). "Significant statistically relationship between the great volcanic eruptions and the count of sunspots from 1610 to the present." *EGU General Assembly Conference Abstracts*.
Caseldine, C. et al. (2005). "Evidence for an extreme climatic event on Achill Island, Co. Mayo, Ireland around 5200–5100 cal. yr BP." *Journal of Quaternary Science* 20(2):169–178.
Catling, D.C., and Zahnle, K.J. (2009). "The escape of planetary atmospheres." *Scientific American* 300:36–43.
Cessford, Craig (2015). "A new dating sequence for Çatalhöyük." *Antiquity* 75:717–725.
Chandler, David (2005). "Birthplace of famous Mars meteorite pinpointed." *New Scientist*.
Cherubini, F. et al. (2011). "CO2 emissions from biomass combustion for bioenergy: atmospheric decay and contribution to global warming." *GCB Bioenergy* 3:413–426.
Chilton, Rodney (2009). *Sudden Cold: An Examination of the Younger Dryas Cold*. First Choice Books.
Choi, Charles (2013). "Asteroid Impact That Killed the Dinosaurs: New Evidence." *Live Science*.
Clube, Victor, and Napier, Bill (1982). *The Cosmic Serpent: A Catastrophist View of Earth History*. Universe Books.
Clube, Victor, and Napier, Bill (1990). *The Cosmic Winter*. B. Blackwell.
Cochrane, E. (2011). "Ancient Testimony for a Comet-like Venus." *Proceedings of the NPA*.
Cole, Stephen (2007). "Global 'Sunscreen' Has Likely Thinned, Report NASA Scientists." NASA.
Collins, Paul (1994). "The Sumerian Goddess Inanna (3400–2200 BC)." *Papers from the Institute of Archaeology* 5:103–118.
Condie, Bill (2015). "Mars lost more water than the volume of the Arctic Ocean." *Cosmos Magazine*.
Condie, Kent (2016). *Earth as an Evolving Planetary System*. Academic Press.
Condron, Alan et al. (2012). "Meltwater routing and the Younger Dryas." *Proceedings of the National Academy of Sciences* 109(49):19928–19933.
Cool Antarctica Editors (2020). "Arctic Hare Facts and Adaptations *Lepus arcticus*." *Cool Antarctica*.
Corfidi, Stephen et al. (2013). "About Derechos." Storm Prediction Center, *NOAA Web Site*.
Cowen, Richard (2000). *History of Life*, Blackwell Science, chapter "The K-T extinction."
Crenson, Matt (2006). "After 10 years, few believe life on Mars." *USAToday*
Cronin, J.R. et al. (1983). "Amino acids in meteorites." *Advances in Space Research* 3(9):5–18.
Cronin, T.M. (2012). "Rapid sea-level rise." *Quaternary Science Reviews* 56:11–30.
Cummings, Vicki et al. (2014). *The Oxford Handbook of the Archaeology and Anthropology of Hunter-Gatherers*. OUP Oxford.
D'Arrigo, R. et al. (2009). "The impact of volcanic forcing on tropical temperatures during the past four centuries." *Nature Geoscience* 2(1):51–56.
D'Huart, Jean et al. (2013). "Hylochoerus meinertzhageni – Forest Hog." In Melletti and Meijaard (eds.), *Mammals of Africa: Volume VI*. Cambridge University Press.
Dalton, Rex (2007). "Archaeology: Blast in the past?" *Nature* 447(7142):256ff.
Davies, Bethan (2020). "Calculating glacier ice volumes and sea level equivalents." *Antarctic Glaciers*.
De Bergh, C. et al. (1991). "Deuterium on Venus: Observations From Earth." *Science* 251(4993):547–549.

De Grazia, A., and Milton, E. (1984). *Solaria Binaria*. Metron Publications.

DeMenocal, Peter (2001). "Cultural Responses to Climate Change During the Late Holocene." *Science* 292(5517):667–673.

Diachenko, Aleksandr et al. (2012). "The gravity model: monitoring the formation and development of the Tripolye culture giant-settlements in Ukraine." *Journal of Archaeological Science* 39(8):2810–2817.

Dietrich, Oliver et al. (2013). "Establishing a Radiocarbon Sequence for Göbekli Tepe. State of Research and New Data." *Neo-Lithics* 1:36–47.

Dlugokencky, E. et al. (2014). "Chemistry of the Atmosphere – Methane." In *Encyclopedia of Atmospheric Sciences*. Academic Press.

Donahue, T. et al. (1997). "Ion/neutral Escape of Hydrogen and Deuterium: Evolution of Water." in *Venus II: Geology, Geophysics, Atmosphere, and Solar Wind Environment*. University of Arizona Press.

Dörr, M. (2011). "Deuterium." in *Encyclopedia of Astrobiology*. Springer.

Doyle, Heather (2020). "What Is Permafrost?" *NASA Climate Kids*.

Dubinin, E. et al. (2017). "The effect of solar wind variations on the escape of oxygen ions from Mars through different channels: MAVEN observations." *Journal of Geophysical Research: Space Physics* 122:11,285–11,301.

Dubinin, Eduard et al. (2013). "Plasma in the near Venus tail: Venus Express observations." *Journal of Geophysical Research: Space Physics* 118(12):7624–7634.

Duma, G. et al. (2003). "Diurnal changes of earthquake activity and geomagnetic Sq-variations." *Natural Hazards and Earth System Sciences* 3(3/4):171–177.

Dykoski, Carolyn (2005). "A high-resolution, absolute-dated Holocene and deglacial Asian monsoon record from Dongge Cave, China." *Earth and Planetary Science Letters* 233(1–2):71–86.

Easterbrook, Don (1999). *Surface Processes and Landforms*. Prentice Hall.

Easterbrook, Don (2019). "Younger Dryas." *Encyclopædia Britannica*.

Easterbrook, Don J. et al. (2011). "Evidence for Synchronous Global Climatic Events." in *Evidence-Based Climate Science*.

Eden, Dan. "Changing Poles" (2011). *Viewzone*. http://viewzone2.com/changingnorth22.html.

Editors of Encyclopaedia Britannica (1998). "Quetzalcoatl, Mesoamerican God." *Encyclopaedia Britannica*.

Editors of the WDC (2020). "Ice cores." World Data Center for Paleoclimatology, NOAA. https://www.ncdc.noaa.gov/data-access/paleoclimatology-data/datasets/ice-core.

Eliade, Mircea (1976). *Occultism, Witchcraft, and Cultural Fashions*. University of Chicago Press.

ESA Editors (2011). "Giotto, ESA's first deep-space mission: 25 years ago." European Space Agency.

Ettinger, Douglas (2016). *The Great Deluge: Fact or Fiction?* Ettinger Journals.

European Space Agency (2019). "From clouds to craters: Mars Express." *Phys.org*.

F. Roos-Barraclough et al. (2002). "A 14 500 year record of the accumulation of atmospheric mercury in peat: volcanic signals, anthropogenic influences and a correlation to bromine accumulation." *Earth and Planetary Science Letters* 202:435–451.

Facts and Details Editors (2016). "Yangshao Culture (5000 B.C. to 3000 B.C.)." *Facts and Details*.

Felix, J. (1912). "Das Mammuth von Boma." *Veröffentlichungen des Städtischen Museum for Völkerkunde zu Leipzig* 4:1–55.

Felix, Robert (2008). *Magnetic Reversals and Evolutionary Leaps: The True Origin of Species*. Sugarhouse Publications.

FERAJ Editors. (2020). "Pi-Puppids 1901-2100: activity predictions." *Feraj.ru*

Ferguson, Charles (1970). "Concepts and Techniques of Dendrochronology." Laboratory of Tree-Ring Research. University of Arizona.

Firestone, R.B. *et al.* (2007). "Evidence for an extraterrestrial impact 12,900 years ago that contributed to the megafaunal extinctions and the Younger Dryas cooling." *Proceedings of the National Academy of Sciences* 104(41):16016-16021.

Firestone, Richard B. (2008). "Impacts, mega-tsunami, and other extraordinary claims." *GSA Today.*

Firestone, Richard B. (2010). "Analysis of the Younger Dryas Impact Layer." *Lawrence Berkeley National Laboratory.*

Firestone, Richard B. *et al.* (2006). *The Cycle of Cosmic Catastrophes: How a Stone-Age Comet Changed the Course of World Culture.* Simon and Schuster.

Forster, Thomas (1829). *Illustrations of the atmospherical origin of epidemic disorders of health.* Meggy & Chalk.

Frechen, M. (2011). "Loess in Europe." *Quaternary Science Journal* 60(1).

Freeborn, B.L. (2014). *The Deep Mystery: The Day the Pole Moved.* Tiw & Elddir Books.

Frey, Karen (2009). "Impacts of Permafrost Degradation on Arctic River Biogeochemistry." *Hydrological Processes* 23:169–182.

Fromm, Michael *et al.* (2000). "Observations of boreal forest fire smoke in the stratosphere by POAM III, SAGE II, and lidar in 1998." *Geophysical Research Letters* 27(9):1407–1410.

Fuller, Errol (2004). *Mammoths: Giants of the Ice Age.* Bunker Hill Publishing, Inc.

Garrick-Bethell, Ian *et al.* (2006). "Evidence for a Past High-Eccentricity Lunar Orbit." *Science* 313(5787):652–655.

Generalic, Eni (2019). "Rare Earth Elements." *Periodni.com.*

Ghosh, Anurag (2010). "What is the Average Asteroid Speed?" *BrightHub website.*

Gill, Prabhjote (2019). "The last major asteroid to hit Earth destroyed 500 square kilometers – and it could have been much worse." *Business Insider.*

Gimbutas, Marija (1956). "The prehistory of Eastern Europe, pt. 1: Mesolithic, Neolithic and copper age cultures in Russia and the Baltic area." *American School of Prehistoric Research.*

Ginenthal, Charles (2015). "Velikovsky's Hydrocarbon Clouds?" in *Carl Sagan and Immanuel Velikovsky.* Lulu Press.

Gladman, B.J. *et al.* (1995). "Ejecta transfer between terrestrial planets." *Proceedings of the 172nd Symposium of the International Astronomical Union.*

Glass, Billy P. (1982). *Introduction to Planetary Geology.* Cambridge University Press.

Gloeckler, G. *et al.* (2000). "Interception of comet Hyakutake's ion tail at a distance of 500 million kilometres." *Nature* 404:576–578.

Gold, Thomas (1979). "Electrical Origin of the Outbursts on Io." *Science* 206(4422):1071–1073.

Goldreich, Peter *et al.* (1966). "Spin-orbit coupling in the solar system." *Astronomical Journal* 71:425.

Gombosil, Tamás (2004). "Thomas Michael Donahue (1921–2004)." *Bulletin of the AAS* 36(4).

Gornitz, Vivien (2013). *Rising Seas: Past, Present, Future.* Columbia University Press.

Gorter, J.D. (1996). "Speculation on the origin of the Bedout high – a large, circular structure of pre-Mesozoic age in the offshore Canning Basin, Western Australia." *PESA News.*

Gounelle, Matthieu *et al.* (2006). "The orbit and atmospheric trajectory of the Orgueil meteorite from historical records." *Meteoritics & Planetary Science* 41:135–150.

Goyer, G. (1960). "Effects of electric fields on water-droplet coalescence." *Journal of Meteorology* 17.

Graham, Russell *et al.* (2016). "Holocene mammoth population extinction." *Proceedings of the National Academy of Sciences* 113(33):9310-9314.

Grebowsky, J.M. *et al.* (1997). "Evidence for Venus lightning." *Venus II.* S.A.

Gribbin, J. *et al.* (1973). "Discontinuous Change in Earth's Spin Rate following Great Solar Storm of August 1972." *Nature* 243.

Grinsted, Aslak (2013). "Relationship between sea level rise and global temperature." *Aslak Grinsted personal website.*

Guevara-Murua, A. et al. (2014). "Observations of a stratospheric aerosol veil from a tropical volcanic eruption in December 1808: is this the Unknown ~1809 eruption?" *Climate of the Past* 10(5):1707–1722.

Guillet, Sebastien et al. (2015). "Toward a more realistic assessment of the climatic impacts of the 1257 eruption." *EGU General Assembly* 17:1268.

Gurnett, D.A. et al. (1991). "Lightning and plasma wave observations from the Galileo flyby of Venus." *Science* 253(5027):1522–1525

Gustavo, Politis et al. (2019). "Campo Laborde: A Late Pleistocene giant ground sloth kill and butchering site in the Pampas." *Science Advances* 5(3).

Guthrie, R. Dale (1990). *Frozen Fauna of the Mammoth Steppe*. The University of Chicago Press.

Haack, H. et al. (2011). "CM Chondrites from Comets? — New Constraints from the Orbit of the Maribo CM Chondrite Fall." *LPI Contributions*.

Hagstrum, J.T. et al. (2017). "Impact-related microspherules in Late Pleistocene Alaskan and Yukon 'muck' deposits signify recurrent episodes of catastrophic emplacement." *Scientific Reports* 7(16620).

Halekas, Jasper et al. (2008). "Lunar Prospector observations of the electrostatic potential of the lunar surface and its response to incident currents." *Journal of Geophysical Research: Space Physics* 113(A9).

Hall, Brenda (2013). "History of the grounded ice sheet in the Ross Sea sector of Antarctica during the Last Glacial Maximum and the last termination." *Geological Society*. Special Publications 381.

Hall, Manly P. (1928). "The Myth of the Dying God." In *The Secret Teachings of All Ages*. Prabhat Prakashan.

Hall, Shannon (2016). "The World's Grandest Canyon May Be Hidden beneath Antarctica." *Scientific American*.

Hammer, C. et al. (1980). "W. Greenland ice sheet evidence of post-glacial volcanism and its climatic impact." *Nature* 288:230–235.

Handwerk, Brian (2005). "Woolly Mammoth DNA Reveals Elephant Family Tree." *National Geographic*.

Hansell et al. (1995). "Optical detection of lightning on Venus." *Icarus* 117:345–351.

Hapgood, Charles (1966). *Maps of the Ancient Sea Kings: Evidence of Advanced Civilization in the Ice Age*. Adventures Unlimited Press.

Hapgood, Charles (1999). *The Path of the Pole*. Adventure Unlimited Press.

Hardy, D.A. (1989). *Thera and the Aegean World. Volume III—Chronology*. Proceedings of the Third International Congress.

Harland, David (2007). *Water and the Search for Life on Mars*, Springer Science & Business Media.

Haynes, Gary (2008). *American Megafaunal Extinctions at the End of the Pleistocene*. Springer.

Hecht, Jeff (1997). "Science: Planet's tail of the unexpected." *New Scientist* 2084.

Heidel, Alexander (1946). *The Gilgamesh Epic and the Old Testament Parallels*. University of Chicago.

Henkel, F.W. (1908). "Halley's Comet." *Popular Astronomy* 15:238–245.

Herdiwijaya, Dhani et al. (2014). "On the Relation between Solar and Global Volcanic Activities." *Proceedings of the 2014 International Conference on Physics*.

Hertz, Heinrich (1900). *Electric Waves: Being Researches on the Propagation of Electric Action with Finite Velocity Through Space*. Macmillan and Company.

Holden, Ted (1993). "Velikovsky Update." Efemeral Research Foundation. http://www.raids.org/skepticfiles/neocat/efr.htm.

Holden, Theodore et al. (2019). *Cosmos In Collision: The Prehistory of Our Solar System and of Modern Man*. Lulu Press, Inc.

Holland, G.J. (1993). "WMO/TC-No. 560, Report No. TCP-31." *World Meteorological Organization*, Geneva, Switzerland.

Hooker, Richard (2020). "General information on the Sumerian Epic Gilgamesh (ca. 2000 B.C.E.)." Arkansas State University.

Hoover, R. (2007). "Microfossils of Cyanobacteria in the Orgueil Carbonaceous Meteorite." NASA.

Hoover, R. (2008). "Comets, Carbonaceous Meteorites, and the Origin of the Biosphere." in *Biosphere Origin and Evolution.* Springer.

Hoover, R. (2011). "Fossils of Cyanobacteria in CI1 Carbonaceous Meteorites: Implications to Life on Comets, Europa, and Enceladus." *Journal of Cosmology* 13.

Hopkins, David, *et al.* (1982). "Paleoecology of Beringia." *New York Academic Press.*

Hopkinson, Deborah (2004). "The Volcano That Shook the world: Krakatoa 1883." *Scholastic.com Storyworks* 11(4):8.

Howard, George (2011). "West to East?: Belgian Tsunami at the Lower Younger Dryas Boundary." *The Cosmic Tusk.*

Howard, George (2020). "Lake Tahoe Comet Tsunami." *The Cosmic Tusk.*

Hoyle, F., and Wickramasinghe, C. (2003). "SARS – a clue to its origins?" *The Lancet* 361(9371):1832.

Hughes, David (2003). "The variation of short-period comet size and decay rate with perihelion distance." *Monthly Notices of the Royal Astronomical Society* 346(2):584–592.

Isachenkov, Vladimir (2015). "Russia to UN: We are claiming 463,000 square miles of the Arctic." *Associated Press.*

Javadinejad, S. *et al.* (2019). "Investigation of monthly and seasonal changes of methane gas with respect to climate change using satellite data." *Applied Water Science* 9(180).

Jha, Vandana (2018). "Sensitivity Studies on the Impact of Dust and Aerosol Pollution Acting as Cloud Nucleating Aerosol on Orographic Precipitation in the Colorado River Basin." *Advances in Meteorology.*

Jia, Xin *et al.* (2017). "Spatial and temporal variations in prehistoric human settlement and their influencing factors on the south bank of the Xar Moron River, Northeastern China." *Frontiers of Earth Science* 11(1):137–147.

Jolly, William Lee (1999). "Hydrogen chemical element." *Encyclopedia Britannica.*

Jull, A.J. *et al.* (1989). "Trends in Carbon-14 Terrestrial Ages of Antarctic Meteorites from Different Sites." *Lunar Planetary Science* 20:488.

Kahng, Sam *et al.* (2012). "Ecology of mesophotic coral reefs – Temperature related depth limits of warm-water corals." *Proceedings of the 12th International Coral Reef Symposium.*

Kaiho, K. *et al.* (2016). "Global climate change driven by soot at the K-Pg boundary as the cause of the mass extinction." *Scientific Reports* 6(28427).

Kaiho, K. *et al.* (2017). "Site of asteroid impact changed the history of life on Earth: the low probability of mass extinction." *Scientific Reports* 7(14855).

Kairiukstis, L. *et al.* (1986). *Methods of Dendrochronology – I. Proceedings of the Task Force Meeting on Methodology of Dendrochronology: East/West Approaches. Krakow, Poland,* IIASA/Polish Academy of Sciences, Warsaw.

Kaminski, J.Y. *et al.* (2004). "A General Framework for Trajectory Triangulation." *Journal of Mathematical Imaging and Vision* 21:27–41.

Kamoun, Paul (1983). "Radar Observations of Cometary Nuclei." Massachusetts Institute of Technology.

Kanevski, Mikhail *et al.* (2011). "Cryostratigraphy of late Pleistocene syngenetic permafrost (Yedoma) in northern Alaska, Itkillik River exposure." *Quaternary Research* 75(3):584–596.

Kennett, Douglas *et al.* (2009). "Shock-synthesized hexagonal diamonds in Younger Dryas boundary sediments." *Proceedings of the National Academy of Sciences* 106(31):12623–12628.

Kerminen, V.-M. *et al.* (2012). "Cloud condensation nuclei production associated with atmospheric nucleation: a synthesis based on existing literature and new results." *Atmospheric Chemistry and Physics* 12(24):12037–12059.

Kher, Aparna (2020). "10 Things You Need to Know about Asteroids." *TimeAndDate.com*

Kiefer, W. S. and Hager, B. H. (1991). "A mantle plume model for the equatorial highlands of Venus." *Journal of Geophysics Research* 96:20947–20966.

King, Hobart (2020). "The Pyroxene Mineral Group." *Geology.com.*

King, Leonard (2007). *Enuma Elish: The seven tablets of creation: The Babylonian and Assyrian legends concerning the creation of the world and of mankind.* Cosimo.

Kinnier-Wilson, James (1979). *The Rebel Lands.* Cambridge University Press.

Kjær, Kurt (2018). "A large impact crater beneath Hiawatha Glacier in northwest Greenland." *Science Advances.*

Klemetti, Erik (2012). "The Mysterious Missing Eruption of 1258 A.D." *Wired.com.*

Kneyda Spins Editor (2013). "Woolly Mammoth Wool?" *Kneyda Spins.*

Knittel, Ulrich (1999). "History of Taal's Activity to 1911 as Described by Fr. Saderra Maso." Institut für Mineralogie und Lagerstättenlehre. RWTH Aachen University.

Kramer, Miriam (2013). "Venus Can Have 'Comet-Like' Atmosphere." *Space.com.*

Krause, Hans (1978). *The Mammoth—In Ice and Snow?,* Stuttgart, Germany, self-published, "Chapter 1: The Berezovka Mammoth."

Kring, David (2020). "Impact induced perturbations of atmospheric sulphur." *Lunar and Planetary Institute.*

Kronk, Gary (1999). *Cometography: Volume 2, 1800–1899: A Catalog of Comets,* Cambridge University Press.

Kronk, Gary (2020). "C/1811 F1 (Great Comet)." *Cometography.com.*

Krzos, Mark (2006). "Frozen Mammoths." https://slideplayer.com/slide/14273917/.

Kunzig, Robert (2013). "Did a Comet Really Kill the Mammoths 12,900 Years Ago?" *National Geographic.*

Kurbatov, A.V. et al. (2006). "A 12,000 year record of explosive volcanism in the Siple Dome Ice Core, West Antarctica." *Journal of Geophysical Research* 111(D12).

Lamb, Hubert F. (1995). *Climate, History, and the Modern World.* London: Routledge.

Lamb, Hubert F. et al. (1995). "Relation between century-scale Holocene and intervals in tropical and temperate zones." *Nature* 373:134.

Lancaster University Editors (2018). "Volcanoes and glaciers combine as powerful methane producers." *Phys.org.*

Langenauer, M. et al. (1993). "Depth-profiles and surface enrichment of the halogens in four Antarctic H5 chondrites and in two non-Antarctic chondrites." *Meteoritics* 28(1):98–104.

Latif, Asad (2009). *Three Sides in Search of a Triangle: Singapore-America-India Relations.* Institute of Southeast Asian Studies.

Lavigne, Franck et al. (2013). "Source of the great A.D. 1257 mystery eruption unveiled, Samalas volcano, Rinjani Volcanic Complex, Indonesia." *Proceedings of the National Academy of Sciences of the United States of America* 110(42).

LaViolette, Paul (2005). *Earth Under Fire: Humanity's Survival of the Ice Age.* Bear & Co.

Lawler, Mark E. et al. (1989). "Iron, magnesium, and silicon in dust from Comet Halley." *Icarus* 80(2):225–242.

Legge, James (1865). *The Chinese Classics Vol. III, Part I.* Trübner & Co.

Leick, Gwendolyn (2002). *A Dictionary of Ancient Near Eastern Mythology.* Routledge.

Lemke, Frank et al. (2014). "The Sumerian K8538 tablet – The great meteor impact devastating Mesopotamia." *ResearchGate.*

Lemmons, Richard (2017). "Goldilocks in space Earth Mars and Venus." *Climate Policy Watcher.*

Leovy, C. E. et al. (1973). "Mechanisms for Mars Dust Storms." *Journal of the Atmospheric Sciences* 30(5):749–762.

Lescaudron, Pierre (2020). "Compelling evidence that SARS-CoV-2 was manmade." *Sott.net.*

Lescaudron, Pierre, with Laura Knight-Jadczyk (2014). *Earth Changes and the Human-Cosmic Connection.* Red Pill Press.

Levermann, Anders. et al. (2013). "The multimillennial sea-level commitment of global warming." *Proceedings of the National Academy of Sciences* 110(34):13745–13750.

Lewin, Roger (1980). "Is your brain really necessary?" *Science* 210(4475):1232–4.

Libal, Angela (2018). "The Temperatures of Outer Space Around the Earth." *Sciencing.com*.

Lin, Weili (2012). "Characteristics and recent trends of sulfur dioxide at urban, rural, and background sites in North China: Effectiveness of control measures." *Journal of Environmental Sciences* 24(1):34–49.

Lippi, Manuela (2010). "The composition of cometary ices as inferred from measured production rates of volatiles." *Carolo-Wilhelmina University*.

Liu, Jean Paul *et al.* (2004). "Reconsidering melt-water pulses 1A and 1B: Global impacts of rapid sea-level rise." *Journal of Ocean University of China* 3:183–190.

Longo, G. *et al.* (1994). "Search for microremnants of the Tunguska cosmic body." *Planetary and Space Science* 42(2):163–177;

Lord, Edana *et al.* (2020). "Pre-extinction Demographic Stability and Genomic Signatures of Adaptation in the Woolly Rhinoceros." *Current Biology*.

Lua, Houyuan *et al.* (2009). "Earliest domestication of common millet (Panicum miliaceum) in East Asia extended to 10,000 years ago." *Proceedings of the National Academy of Sciences* 106(18):7367–7372.

Lunar and Planetary Institute Editors (2014). "How could ALH84001 get from Mars to Earth?" *Lunar and Planetary Institute*.

Lunar and Planetary Institute Editors (2020). "Shaping the Planets: Impact Cratering." *Lunar and Planetary Institute*.

Luzum, Brian *et al.* (2011). "The IAU 2009 system of astronomical constants: The report of the IAU working group on numerical standards for Fundamental Astronomy." *Celestial Mechanics and Dynamical Astronomy* 110(4):293–304.

Macdonald, Fiona (2019). "Ancient Carvings Show Evidence of a Comet Swarm Hitting Earth Around 13,000 Years Ago." *Science Alert*.

Manning, S.W. (1992). "Thera, Sulphur and Climatic Anomalies." *Oxford Journal of Archaeology* 11:245–253.

Marchitelli, V. *et al.* (2020). "On the correlation between solar activity and large earthquakes worldwide". *Scientific Reports* 10(11495).

Marcia, Malory (2013). "New evidence that cosmic impact caused Younger Dryas extinctions." *Phys.org*.

Mariner Stanford Group (1967). "Venus: ionosphere and atmosphere as measured by dual-frequency radio occultation of Mariner V." *Science* 158:1678–1683.

Market, Patrick (2011). "A Record Setting Temperature Shift 100 Years Ago Today." *CBS Saint Louis*.

Martini, Peter *et al.* (2010). *Landscapes and Societies: Selected Cases.* Springer Science & Business Media.

Mason, Brian *et al.* (1970). "Minor and Trace Elements in Meteoritic Minerals." *Smithsonian Contributions to the Earth Sciences*: 1–17.

Masse, W. *et al.* (2007). "Missing in Action? Evaluating the Putative Absence of Impacts by Large Asteroids and Comets during the Quaternary Period." *ResearchGate*.

Mather, T., Pyle, D., and Oppenheimer, C. (2013). "Tropospheric Volcanic Aerosol." In *Volcanism and the Earth's Atmosphere* (Ed. A. Robock and C. Oppenheimer).

Mayewski, P.A. *et al.* (1993). "The Atmosphere During the Younger Dryas." *Science* 261:195–197.

McAneney, Jonny *et al.* (2019). "Absolute tree-ring dates for the Late Bronze Age eruptions of Aniakchak and Thera in light of a proposed revision of ice-core chronologies." *Antiquity* 93(367):99–112.

McDonough, W.F. *et al.* (1995). "The composition of the Earth." *Chemical Geology* 120(3–4):223–253.

McFadden, Lucy-Ann *et al.* (2006). *Encyclopedia of the Solar System*. Elsevier.

Meehan, Richard (1999). "Oak." In *Ignatius Donnelly and the end of the world*. Kirribili Press.

Meier, M.M. *et al.* (2016). "Mercury (Hg) in meteorites: Variations in abundance, thermal release profile, mass-dependent and mass-independent isotopic fractionation." *Geochimica et Cosmochimica Acta* 182:55–72.

Message To Eagle Editors (2017). "The Knap Of Howar: One Of The Oldest And Well-Preserved Neolithic Complexes Orkney, Scotland." *Message To Eagle.*

Milbrath, Susan (1999). *Star Gods of the Maya: Astronomy in Art, Folklore, and Calendars.* University of Texas Press.

Miller, Gifford H. *et al.* (2012). "Abrupt onset of the Little Ice Age triggered by volcanism and sustained by sea-ice/ocean feedbacks." *Geophysical Research Letters* 39(2).

Mingren, Wu (2017). "The Descent of Inanna into the Underworld: A 5,500-Year-Old Literary Masterpiece." *Ancient Origins.*

Mitchell, Don (2003). "First Pictures of the Surface of Venus." *Mentallandscape.com.*

Mittelholz, Anna *et al.* (2020). "Timing of the martian dynamo: New constraints for a core field at 4.5 and 3.7 Ga." *Science Advances.*

Moelling, K. and Broecker, F. (2019). "Viruses and Evolution – Viruses First? A Personal Perspective." *Frontiers in microbiology* 10:523.

Mollanen, Jamo (2009). "Impact structures of the World." *Somerikko.net.*

Moore, Christopher *et al.* (2017). "Widespread platinum anomaly documented at the Younger Dryas onset in North American sedimentary sequences." *Scientific Reports* 7(44031).

Morris, John (1993). "Did the Frozen Mammoths Die in the Flood or in the Ice Age?" *Institute For Creation Research.*

MSO Editors (2007). "Pi Puppids." *Meteor Showers Online.*

Müller, Johannes *et al.* (2016). *Trypillia Mega-Sites and European Prehistory: 4100–3400 BCE.* Taylor & Francis.

Narcisi, Biancamaria *et al.* (2019). "Multiple sources for tephra from AD 1259 volcanic signal in Antarctic ice cores." *Quaternary Science Reviews* 210:164–174.

Narlikar, J.V. *et al.* (2003). "Balloon experiment to detect micro-organisms in the outer space." *Astrophysics and Space Science* 285(2):555–62.

NASA Content Administrator (2008). "Io: The Prometheus Plume." *Nasa.gov.*

NASA Content Administrator (2008). "Valles Marineris: The Grand Canyon of Mars." *Nasa.org.*

NASA Content Administrator (2012). "Mariner 2 Data from Venus." *Nasa.com.*

NASA Editors (2005). "A Record from the Deep: Fossil Chemistry." *NASA Earth Observatory.*

NASA Editors (2019). "In Depth: Meteors & Meteorites." NASA Science – Solar system exploration.

Neuville, Henri (1919). "On the Extinction of the Mammoth." *Annual Report of the Smithsonian Institution.*

Newhall, Christopher *et al.* (1982). "The Volcanic Explosivity Index (VEI): An Estimate of Explosive Magnitude for Historical Volcanism." *Journal of Geophysical Research* 87(C2):1231–1238.

NOAA Editors (2017). "On This Day: Historic Krakatau Eruption of 1883." *NOAA – National Centers for Environmental Information.*

Noonan, John W. *et al.* (2018). "Ultraviolet Observations of Coronal Mass Ejection Impact on Comet 67P/Churyumov by Rosetta Alice." *The Astronomical Journal* 156:16.

Nuttall, M. (2005). *Encyclopedia of the Arctic.* New York: Routledge Publishing.

Oard, Michael (1979). "A Rapid Post-Flood Ice Age." *Creation Research Society Quarterly* 16:29–37.

Odintsov, S. *et al.* (2006). "Long-period trends in global seismic and geomagnetic activity and their relation to solar activity." *Phys. Chem. Earth* 31(1–3):88–93.

Omerbashich, Mensur (2012). "Astronomical alignments as the cause of ~M6+ seismicity." *arXiv*:1104.2036v7.

Onoue, Tetsuji *et al.* (2012). "Deep-sea record of impact apparently unrelated to mass extinction in the Late Triassic." *Proceedings of the National Academy of Sciences* 109(47):19134-9.

Oppenheimer, Clive (2003). "Ice core and palaeoclimatic evidence for the timing and nature of the great mid-13th century volcanic eruption." *International Journal of Climatology* 23:417–426.

ORCID Editors (2020). "Mensur Omerbashich Biography." *Orcid.org.*

Oskina, N.S. *et al.* (2019). "Warm-Water Planktonic Foraminifera in Kara Sea Sediments." *Oceanology* 59:440–450.

Ostenso, Ned (1998). "Arctic Ocean." *Encyclopædia Britannica.*

Oyama, V.I. *et al.* (1979). "Venus lower atmospheric composition: analysis by gas chromatography." *Science* 203(4382):802–805.

Park, Ryan S. (2007). "JPL Small-Body Database – 26P/Grigg-Skjellerup." Jet Propulsion Laboratory. NASA.

Pätzold, M. *et al.* (2007). "The structure of Venus' middle atmosphere and ionosphere." *Nature* 450:657–660.

Pearce, N.J.G. *et al.* (2004). "Identification of Aniakchak (Alaska) tephra in Greenland ice core challenges the 1645 BC date for Minoan eruption of Santorini." *Geochemistry, Geophysics, Geosystems* 5(3).

Penglase, Charles (2003). *Greek Myths and Mesopotamia: Parallels and Influence in the Homeric Hymns and Hesiod.* Routledge.

Peratt, A. *et al.* (1988). "Filamentation of Volcanic Plumes on Io." *Astrophysics and Space Science* 144:451–456.

Petaev, Michail *et al.* (2013). "Large Pt anomaly in the Greenland ice core points to a cataclysm at the onset of Younger Dryas." *Proceedings of the National Academy of Sciences* 110:12917–12920.

Pettit, Paul, and White, Mark (2012). *The British Palaeolithic: Human Societies at the Edge of the Pleistocene World.* Abingdon, UK: Routledge.

Pham, Q.T. (2014). *Food Freezing and Thawing Calculations.* Springer.

Phillips, Tony (2003). "Approaching Mars." *Nasa.gov.*

Pierazzo, Elizabetta *et al.* (2010). "Ozone perturbation from medium-size asteroid impacts in the ocean." *Earth and Planetary Science Letters* 299(3–4):263–272.

Pino, Mario *et al.* (2019). "Sedimentary record from Patagonia, southern Chile supports cosmic-impact triggering of biomass burning, climate change, and megafaunal extinctions at 12.8 ka." *Scientific Reports* 9(4413):13.

PM News Nigeria Editors (2020). "Akeredolu lied, Akure blast caused by meteorite: Professor Adepelumi." *PM News Nigeria.*

Pollock, S. (1992). "Bureaucrats and managers, peasants and pastoralists, imperialists and traders: Research on the Uruk and Jemdet Nasr periods in Mesopotamia." *Journal of World Prehistory* 6:297–336.

Port, Jake (2016). "What causes an aurora over the poles?" *Cosmos Magazine.*

Potter, R.W.K. *et al.* (2013). "Numerical modeling of asteroid survivability and possible scenarios for the Morokweng crater-forming impact." *Meteoritics & Planetary Science* 48:744–757.

Prager, Christoph *et al.* (2009). "Geological controls on slope deformations in the köfels rockslide area (Tyrol, Austria)." *Austrian Journal of Earth Sciences* 102:4–19.

Proust, Christine (2009). "Numerical and Metrological Graphemes: From Cuneiform to Transliteration." *Cuneiform Digital Library Journal.*

Prouty, W. F. (1952). "Carolina Bays and their Origin." *Geological Society of America Bulletin* 63(2):167–224.

Pruitt, Sarah (2017). "Scientists Say They Could Bring Back Woolly Mammoths Within Two Years." *History.com.*

Pure Insight Editors (2003). "Reflections on History: Natural Disasters and the Decline and Fall of the Xia Dynasty." *PureInsight.com*

Rajmon, David (2019). "Suspected Earth Impact Sites database." *David.rajmon.cz.*

Rascovan, N. *et al.* (2019). "Emergence and Spread of Basal Lineages of *Yersinia pestis* during the Neolithic Decline." *Cell* 176(1–2):295–305.e10.

Raymond, Sean (2014). "Real-life sci-fi worlds #3: the oscillating Earth." *Planetplanet.*

Reid, Anthony (2016). "Revisiting Southeast Asian History with Geology: Some Demographic Consequences of a Dangerous Environment." In Greg Bankoff *et al.*, *Natural Hazards and Peoples in the Indian Ocean World*. Palgrave Macmillan.

Rhawn, Joseph *et al.* (2010). "Comets and Contagion: Evolution, Plague, and Diseases From Space." *Journal of Cosmology* 7:1750–1770.

Robbins, S. J. *et al.* (2012). "A new global database of Mars impact craters ≥1 km: 2. Global crater properties and regional variations of the simple-to-complex transition diameter." *Journal of Geophysical Research* 117(E6).

Rose, L.E. *et al.* (1974). "Velikovsky and the sequence of planetary orbit." *PENSEE Journal VIII*. Mikamar Publishing.

Roux, Georges (1992). *Ancient Iraq*. Penguin UK.

Rowe, Harry *et al.* (2004). "Hydrologic-energy balance constraints on the Holocene lake-level history of lake Titicaca, South America." *Climate Dynamics* 23:439–454.

Royal Society's Krakatoa Committee (1888). *The Eruption of Krakatoa: And Subsequent Phenomena*. Trübner & Company.

Ruan, Jiaping *et al.* (2015). "A high resolution record of sea surface temperature in southern Okinawa Trough for the past 15,000 years." *Palaeogeography, Palaeoclimatology, Palaeoecology* 426:209–215.

Ruddiman, W.F. *et al.* (1981). "The mode and mechanism of the last deglaciation: Oceanic evidence." *Quaternary Research*.

Ryan, Frank (2013). *Virolution*. Harper Collins UK.

Salvatore, M.R. *et al.* (2015). "On the origin of the Vastitas Borealis Formation in Chryse and Acidalia Planitiae, Mars." *JGR Planets* 119:2437–2456.

Sanderson, Ivan (1960). "Riddle of the Frozen Giants." *Saturday Evening Post*.

Satake, K. *et al.* (2007). "Long-Term Perspectives on Giant Earthquakes and Tsunamis at Subduction Zones." *Annual Review of Earth and Planetary Sciences* 35(1):355.

Schultz, Colin (2013). "Scientists Just Found a Woolly Mammoth That Still Had Liquid Blood." *Smithsonian Magazine*.

Scott, Michon *et al.* (2016). "Sea Ice." *NASA Earth Observatory*.

Scranton, Laird (2012). *The Velikovsky Heresies: Worlds in Collision and Ancient Catastrophes Revisited*. Simon and Schuster.

Seifert, Joachim, and Lemke, Frank (2016a). "Climate Pattern Recognition In The Mid-To-Late Holocene (4800 BC To 2800 BC, Part 3)." *ResearchGate*.

Seifert, Joachim, and Lemke, Frank (2016b). "Climate Pattern Recognition In The Mid-To-Late Holocene (2900 BC To 1650 BC, Part 4)." *ResearchGate*.

Seifert, Joachim, and Lemke, Frank. (2019). "The Sumerian K8538 tablet. The great meteor impact devastating Mesopotamia – a 2019 translation addendum." *ResearchGate*.

Seki, Arisa *et al.* (2012). "Mid-Holocene sea-surface temperature reconstruction using fossil corals from Kume Island, Ryukyu, Japan." *Geochemical Journal* 46(3).

Senthil, Kumar *et al.* (2008). "Impact fracturing and structural modification of sedimentary rocks at Meteor Crater, Arizona." *Journal Of Geophysical Research* 113:E09009.

Serra, R. *et al.* (1994). "Experimental hints on the fragmentation of the Tunguska cosmic body." *Planetary and Space Science* 42(9):777–783.

Serreze, M.C. *et al.* (2000). "Representation of Mean Arctic Precipitation from NCEP–NCAR and ERA Reanalyses." *Journal of Climate* 13(1):182–201.

Serreze, M.C., and Barry, R.G. (2005). *The Arctic Climate System*. Cambridge University Press.

Settegast, Mary (1987). *Plato Prehistorian*. The Rotenberg Press.

Severinghaus, Jeffrey *et al.* (1998). "Timing of abrupt climate change at the end of the Younger Dryas interval from thermally fractionated gases in polar ice." *Nature* 391 (6663):141–146.

Shan, Jie *et al.* (2018). *Topographic Laser Ranging and Scanning: Principles and Processing*. Second edition. CRC Press.

Sharp, Tim (2012). "How Big is Mars? Size of Planet Mars." *Space.com*.

Shelach, Gideon (2000). "The Earliest Neolithic Cultures of Northeast China: Recent Discoveries and New Perspectives on the Beginning of Agriculture." *Journal of World Prehistory* 14:363–413.

Sheldrake, Rupert (2009). *Morphic Resonance: The Nature of Formative Causation*. Inner Traditions/Bear.

Sherman, Howard (2014). *Mythology for Storytellers: Themes and Tales from Around the World*. Routledge.

Shestopalov, I.P. et al. (2014). "Relationship between solar activity and global seismicity and neutrons of terrestrial origin." *Russian Journal of Earth Sciences* 14(ES1002).

Shivaji, S. et al. (2009). "Janibacter hoylei sp. nov., Bacillus isronensis sp. nov. and Bacillus aryabhattai sp. nov., isolated from cryotubes used for collecting air from upper atmosphere." *International Journal of Systematic and Evolutionary Microbiology* 59(Pt 12):2977–86.

Shu, Catherine (2011). "A shift in time." *Taipei Times*.

Shumskiy, P.A. (1959). "Is Antarctica a Continent or an Archipelago?" *Journal of Glaciology* 3(26):455–457.

Shuvalov, V. (2009). "Atmospheric erosion induced by oblique impacts." *Meteoritics & Planetary Science* 44(8):1095–1105.

Siegel, Ethan (2014). "Does water freeze or boil in space?" *Medium.com*.

Silvia, Phillip et al. (2018). "The 3.7kaBP Middle Ghor Event: Catastrophic Termination of a Bronze Age Civilization." Conference: Annual Meeting of the American Schools of Oriental Research (ASOR).

Simmons, William et al. (2019). "Olivine." *Encyclopædia Britannica*.

Sitchin, Zecharia (1976). *The 12th Planet*. Simon and Schuster.

Slatyer, Will (2014). *Life/Death Rhythms of Ancient Empires – Climatic Cycles Influence Rule of Dynasties*. Partridge India.

Smith, Stephen (2020). "Langmuir Sheaths." *Thunderbolts.info*.

Sonik, Karen (2008). "Bad King, False King, True King: Apsû and His Heirs." *Journal of the American Oriental Society* 128(4):737–743.

Space.com Editors (2012). "Mars Surface Scarred by 635,000 Big Impact Craters." *Space.com*.

Spataro, Michela et al. (2010). "Centralisation or Regional Identity in the Halaf Period? Examining Interactions within Fine Painted Ware Production." *Paléorient* 36(2):91–116.

Stankowski, Wojciech (2011). "Luminescence and radiocarbon dating as tools for the recognition of extraterrestrial impacts." *Geochronometria* 38(1):50–54.

Stankowski, Wojciech et al. (2007). "Luminescence Dating of the Morasko (Poland), Kaali, Ilumetsa and Tsõõrikmäe (Estonia) Meteorite Craters." *Geochronometria* 28(1):25–29.

Starkweather, Llan (2007). "Earth Without Polarity." *Lulu.com*.

Steffensen, J.P. (2020). "Data, icesamples and software." Centre for Ice and Climate. University of Copenhagen. http://www.iceandclimate.nbi.ku.dk/data/.

Stewart, John Massey (1977). "Frozen Mammoths from Siberia Bring the Ice Ages to Vivid Life." *Smithsonian*.

Stich, Adrienne et al. (2012). "Soot as Evidence for Widespread Fires at the Younger Dryas Onset (YDB, 12.9 ka)." *Quaternary International* 279–280:468.

Stuart, Anthony et al. (2011). "Extinction chronology of the cave lion Panthera spelaean." *Quaternary Science Reviews* 30:2329–2340.

Stuut, J-B.W. et al. (2014). "The significance of particle size of long-range transported mineral dust." *PAGES News* 22(2):14–15.

Sukachev, V.N. (1914). "The study of plant remains from the food mammoth found on the Berezovka River." Nauchnye rezul'taty ekspeditsii, *snaryazhennoi Akademiyei Nauk dlya raskopki mamonta. naidennogo po r. Berezovke* 1901(3):1e17.

Sverdrup, H.U. *et al.* (1942). *The Oceans, Their Physics, Chemistry, and General Biology.* Prentice-Hall.

Sweatman, Martin (2018). *Prehistory Decoded.* Troubador Publishing.

Takashimizu, Yasuhiro *et al.* (2016). "Reworked tsunami deposits by bottom currents: Circumstantial evidences from Late Pleistocene to Early Holocene in the Gulf of Cádiz." *Marine Geology* 377:95–109.

Talbott, David *et al.* (2004). "Electric Jets on Io." *Thunderbolts.info.*

Talbott, David *et al.* (2006). "Deep Impact and Shoemaker-Levy 9." *Thunderbolts.info.*

Taylor, W. *et al.* (1979). "Evidence for lighting on Venus." *Nature* 279:614–616.

Terada, Kentaro *et al.* (2017). "Biogenic oxygen from Earth transported to the Moon by a wind of magnetospheric ions." *Nature Astronomy.*

The Independent Editors (2006). "How Krakatoa made the biggest bang." *The Independent.*

The Watchers Editors (2015). "Cosmic-solar radiation as the cause of earthquakes and volcanic eruptions." *The Watchers.*

Thiele, Edwin R. (1983). *The Mysterious Numbers of the Hebrew Kings.* Kregel Academic.

Thompson, L.G. *et al.* (1995). "Late Glacial Stage and Holocene Tropical Ice Core Records from Huascarán. Peru." *Science*

Thompson, Robert *et al.* (1980). "Shasta ground sloth (*Nothrotheriops shastensis hoffstetter*) at Shelter Cave, New Mexico: Environment, diet, and extinction." *Quaternary Research* 14(3):360–376.

Thompson, Thomas L. (2000). *Early History of the Israelite People: From the Written & Archaeological Sources.* Brill.

Thompson, William Irwin (1996). *The Time Falling Bodies Take To Light: Mythology, Sexuality and the Origins of Culture.* Palgrave Macmillan.

Thornhill, W. (2001). "Mars and the Grand Canyon." *Holoscience.com.*

Thornhill, W. (2005). "Comet Tempel 1's Electrifying Impact." *Holoscience.com.*

Thornhill, W. (2005). "The Electric Universe: Part II. Discharges and Scars." *Thunderbolts.info.*

Thornhill, W. *et al.* (2007). *The Electric Universe.* Mikamar Publishing.

Treiman, A.H. (2005). "The nakhlite meteorites: Augite-rich igneous rocks from Mars." *Geochemistry* 65(3):203–270.

Trigo-Rodríguez, Josep M. *et al.* (2019). "Accretion of Water in Carbonaceous Chondrites: Current Evidence and Implications for the Delivery of Water to Early Earth." *Space Science Reviews.*

Turner, J. *et al.* (2009). "Record low surface air temperature at Vostok station, Antarctica." *Journal of Geophysical Research* 114:D24.

Tuttle, Robert J. (2012). *The Fourth Source: Effects of Natural Nuclear Reactors.* Universal Publishers.

Ukraintseva, Valentina (2013). *Mammoths and the Environment.* Cambridge University Press.

University of Florida discussion board (2012). "Freezing Tissue." *Biotech.ufl.edu.*

Van Helden, Albert (2009). "The beginnings, from Lipperhey to Huygens and Cassini." *Experimental Astronomy* 25 (1–3).

Vanneste, H. *et al.* (2016). "Elevated dust deposition in Tierra del Fuego (Chile) resulting from Neoglacial Darwin Cordillera glacier fluctuations." *Journal of Quaternary Science* 31:713–722.

Vasilyev, N.V. (1999). "Ecological consequences of the Tunguska catastrophe." *in Problemi radioekologii i pogranichnikh discipline.*

Vaughan, David *et al.* (1999). "Reassessment of Net Surface Mass Balance in Antarctica." *Journal of Climate* 12(4):933–946.

Velikovsky, Immanuel (1950). *Worlds in Collision.* Macmillan Publishers.

Venzke, E. (2013). "Global Volcanism Program, volcanoes of the World." Smithsonian Institution.

Vernet, Robert *et al.* (2000). "Isotopic Chronology of the Sahara and the Sahel During the Late Pleistocene and the Early and Mid-Holocene (15,000–6000 BP)." *Quaternary International* 68:385–387.

Villanueva, G.L. *et al.* (2015). "Strong water isotopic anomalies in the Martian atmosphere: Probing current and ancient reservoirs." *Science* 348(6231).

Vogel, J.S. *et al.* (1990a). "Vesuvius/Avellino, one possible source of seventeenth century BC climatic disturbances." *Nature* 344:534–537.

Vogel, J.S. *et al.* (1990b). "Letters to Nature: Vesuvius/Avellino, one possible source of seventeenth century BC climatic disturbances." *Nature* 344(6266):534–537.

Wall, Mike (2010). "Venus' Atmosphere Proves a Real Drag, Leading to a Discovery." *Space.com*.

Wall, Mike (2011). "Did Comet Cause Solar Explosion? Hardly, Experts Say." *Space.com*.

Wall, Mike (2015). "Mars Lost Atmosphere to Space as Life Took Hold on Earth." *Space.com*.

Wallace-Hadrill, Andrew (2010). "Pompeii: Portents of Disaster." *BBC History*.

Wallis, J. *et al.* (2013). "The Polonnaruwa meteorite: oxygen isotope, crystalline and biological composition." *Journal of Cosmology* 22(2).

Ward, Dale (2016). "The climate of the Holocene." *University of Arizona*.

Watts, Anthony (2012)."The Intriguing Problem Of The Younger Dryas—What Does It Mean And What Caused It?" *Wattsupwiththat.com*.

Weissman, Paul (2020). "Comet – The modern era." *Encyclopedia Britannica*.

Wikipedia Editors (2002). "Escape Velocity." *Wikipedia*.

Wikipedia Editors (2004). "Clarence Birdseye." *Wikipedia*.

Wikipedia Editors (2006). "Bamboo Annals." *Wikipedia*.

Wikipedia Editors (2006). "Inanna." *Wikipedia*.

Wikipedia Editors (2006). "Worlds In Collision." *Wikipedia*.

Wikipedia Editors (2007). "Abundance of elements in Earth's crust." *Wikipedia*.

Wikipedia Editors (2007). "The Piora Oscillation." *Wikipedia*.

Wikipedia Editors (2017). "Chertovy Vorota Cave." *Wikipedia*.

Wikipedia Editors (2020). "List of Natural satellites." *Wikipedia*.

Wilford, John Noble (1978). "Argon Level on Venus Stirs Debate." *New York Times*.

Wilson, H.C. (1908). "The Next Apparition of Halley's Comet." *Popular Astronomy* 16:265–270.

Witze, Alexandra (2008). "Climate change: Losing Greenland." *Nature* 452:798–802.

Witze, Alexandra (2012). "Earth: Volcanic bromine destroyed ozone: Blasts emitted gas that erodes protective atmospheric layer." *Science News* 182(1):12.

Witzel, Michael (2012). *The Origins of the World's Mythologies*. Oxford University Press.

Wolbach, Wendy *et al.* (2018). "Extraordinary Biomass-Burning Episode and Impact Winter Triggered by the Younger Dryas Cosmic Impact ~12,800 Years Ago." *The Journal of Geology* 126(2).

Wright, Frederick G. (1902). *Asiatic Russia*. McClure, Phillips and Co.

Xiao, Jule *et al.* (2004). "Holocene vegetation variation in the Daihai Lake region of north-central China: a direct indication of the Asian monsoon climatic history." *Quaternary Science Reviews* 23(14–15):1669–1679.

Xue, Jibin *et al.* (2009). "A new high-resolution Late Glacial-Holocene climatic record from eastern Nanling Mountains in South China." *Chinese Geographical Science* 19:274–282.

Zielinski, G.A. (1994). "Record of Volcanism Since 7000 B.C. from the GISP2 Greenland Ice Core and Implications for the Volcano-Climate System." *Science New Series* 264(5161).

Zielinski, G.A. *et al.* (1996). "A 110,000-Yr Record of Explosive Volcanism from the GISP2 (Greenland) Ice Core." *Quaternary Research* 45(2):109–118.

Zijlstra, Albert (2016). "1809: The missing volcano." *Volcano Café*.

Zimov, Sergey *et al.* (2014). "Role of Megafauna and Frozen Soil in the Atmospheric CH4 Dynamics." *Public Library of Science*.

Zukerman, Wendy (2011). "Warmer oceans release CO2 faster than thought." *New Scientist*.

Zwolinski, Zbigniew *et al.* (2017). "Existing and Proposed Urban Geosites Values Resulting from Geodiversity of Poznań City." *Quaestiones Geographicae* 36(3):125–149.

Index

ablation, i, 29, 32, 33, 54, 125
accretion, 55, 95, 99
aerosol, 26, 135, 149, 158, 160, 161, 199
albedo, 54, 55
Annals, 135, 137
Antarctica, 4, 7, 25, 43, 44, 89–95, 105, 107–109, 112, 120, 121, 136, 141, 147–149, 160, 195, 224
Arctic, 2, 4, 8, 20, 37, 42–44, 46, 49, 62, 69, 71, 91, 93–95, 110, 219, 225
argon, 101, 216
Assyria, 212, 215
asteroid, 12, 13, 15, 17, 26, 29, 30, 46, 47, 75, 89, 95, 100, 103, 123, 126, 150–152, 178, 214, 224, 227
atmosphere, i, 19, 26, 27, 29, 30, 32–34, 50, 53–56, 65, 71, 73, 75, 100, 102, 119, 120, 122, 123, 125, 126, 131, 133, 135, 139, 141, 149, 151, 152, 157–162, 165, 170, 173, 179–181, 191, 193, 195, 197–199, 203, 223, 224, 233
axis, 37, 44, 46

bedrock, 41, 90, 91, 93, 95, 108, 125, 222
Berezovka, 21, 49, 51, 53
blizzard, 4
boulder, 54, 57, 222
bromide, 151, 152
buttercups, 22

calcium, 89, 122, 123
canyon, 85, 87, 91
carbon, 11, 12, 15, 21, 37, 50, 60, 63, 89, 95, 105, 112, 120–123, 132, 135, 141, 144, 149, 158, 180, 181, 185, 195, 201, 203, 207–209, 223, 224
carbon dioxide, 21, 120, 121, 123, 180, 181, 185, 203
Cariaco, 119
Carolina Bays, 32
catastrophe, iii, 1, 2, 17, 63, 65, 95, 99, 114, 115, 137, 139, 152, 173, 174, 212, 213, 227, 230
cave, 2, 4, 123, 145, 147, 148, 153, 186, 200, 201
Chelyabinsk, 53, 227, 228
chevron, 59, 60, 86, 224
Chicxulub, 15, 122, 123, 230
China, 37, 119, 126, 135, 137, 145, 147, 148, 180, 207, 208, 224

clay, 207, 215, 216, 230
Clovis, 8, 9, 11–13, 15, 65
cold, 3–5, 7, 12, 21, 25, 34, 37, 94, 120, 159, 174, 187, 207
collapse, 87, 98, 114, 144, 145, 203, 205–208
combustion, 120, 149
comet, iii, iv, 1, 7, 11–13, 15, 17, 19–21, 26, 29–32, 44, 46, 51, 53–56, 59, 60, 63, 65, 69, 77, 98–103, 105, 112, 113, 117, 118, 121–130, 135, 137, 139, 141, 144, 148, 158, 159, 161–163, 165, 168–170, 172–174, 177, 178, 180–183, 192, 195, 197, 203, 208, 209, 211–214, 220, 222, 224, 227–231, 233, 236
condensation, 54, 55, 199
cooling, iii, 7, 12, 21, 25, 26, 34, 47, 54, 55, 65, 68, 69, 99, 105, 109–112, 118–120, 126, 157, 174, 185, 187, 195, 198, 205, 208, 220, 222, 225
coral, 37, 38, 43, 110, 114, 126
crater, 12, 13, 15, 17, 19, 20, 32, 56, 60, 74–76, 83, 87, 100, 124–126, 141, 150, 153, 158, 159, 223–225, 228
Cretaceous, 13, 15, 230
crust, 7, 13, 37, 45–47, 54, 73, 99, 151, 152, 171, 172
crystal, 14, 22, 126, 209, 222

dating, 7, 13, 60, 105, 112, 114, 125, 132, 134, 144, 157, 158, 207, 208, 215, 216, 223–225
deluge, 54, 69, 79, 80, 113, 115
dendrochronology, 134, 187
density, 29, 30, 61, 74, 165, 166, 171
derecho, 26
deuterium, 180, 182, 191, 192, 203
diamond, 11, 13–15, 158
dinosaur, 13, 15, 123, 230
discharge, 67, 69, 77–81, 83, 85, 87, 89, 91, 97, 103, 113, 114, 118, 127, 128, 169, 170
dust, 13, 21, 34, 53–56, 75, 126, 133, 135, 152, 158, 159, 162, 163, 165, 173, 181, 182, 187, 197–199, 203

earthquake, iii, 57, 60, 117, 124, 137, 165, 166, 168–170, 172, 197
Egypt, 137, 206, 208, 211
ejecta, 17, 18, 32, 33, 56, 60, 122, 126, 179

electric, 55, 56, 67, 77–81, 83, 85, 87, 89, 91, 95, 97, 98, 101–103, 113, 114, 118, 127, 128, 166, 169–173, 191
equator, 37, 99, 120, 172
erosion, 19, 29, 30, 51, 56, 75, 85–87
eruption, 26, 37, 38, 54, 56, 57, 132–135, 141, 143, 144, 154–161, 163, 165–170, 172, 174, 180, 181, 193, 194, 197, 223

fauna, 8, 12, 34, 43, 44, 50, 119, 123, 152, 153
fire, 26, 29, 49, 152, 165, 202, 212
Firestone, Richard B., 12, 15, 17–20, 30, 46, 59
flare, 77, 166, 169–171
flash-freezing, 2, 20–22, 26, 53
flood, 22, 32, 51, 59, 85, 114, 115, 171, 201
flora, 34, 43, 44, 119
fluted point, 8, 9
foramanifera, 43
fossil, 43, 44, 209, 230
fractal, 81, 235
Fragment, 17, 19, 20, 30–32, 46, 53, 56, 59, 77, 89, 118, 163, 181, 194, 200, 214, 223, 225, 230
Fullerene, 11, 12, 15, 158
fur, 4

geochemical, 43, 133, 155, 157, 158, 202
geographic pole, 34, 37–39, 41, 43, 44, 46
glacier, 41, 71, 111, 119, 125, 185, 187, 197, 198, 200
glass, 12, 15, 101, 133, 134, 158
Gobekli Tepe, 63
Greenland, 7, 13, 42, 47, 91, 93, 94, 118–122, 124, 125, 133–136, 141, 148, 186, 187, 192, 193, 223, 225

Halley's Comet, 161–163, 220, 227
Hapgood, Charles Hutchins, 7, 37, 38, 42, 46, 53, 105, 107
heliosphere, 169
helium, 12, 15
hemisphere, 7, 11, 35, 42, 56, 68, 74, 75, 120, 124, 125, 132, 134, 135, 148, 151, 157, 219, 225
Holocene, 7, 8, 43, 60, 119, 128, 134, 147, 173, 187, 195, 198, 201, 203, 206, 208, 221, 222
Hudson, 18, 30, 41–44, 59
hurricane, 54, 55, 57
hydrocarbon, 180, 181

ice, 13, 22, 49–51, 93, 105, 108, 111, 121, 141, 157, 186, 195, 198, 207, 219, 225
ice cap, 41, 42, 59, 68, 94, 185
ice core, 7, 13, 94, 110, 120–122, 131–135, 139, 141, 143, 144, 147–149, 152, 155, 158–160, 162, 179, 183, 186, 192–195, 198, 219, 223, 225
impact, iii, 7, 12, 13, 15, 17–20, 26, 29–33, 39, 41, 46, 49, 53, 54, 56, 59, 60, 69, 75, 77, 100, 122–126, 128, 135, 141, 148–152, 157, 159, 166, 168, 173, 181, 193, 197, 199, 205, 221–225, 228, 230, 231, 233
Inanna, 178, 211–216, 220
ionosphere, 55, 56, 102, 166, 170
iridium, 13, 15, 158
isotope, 5, 12, 15, 100, 110, 119, 131, 162, 191, 209

Japan, 43, 80, 120, 168
Jupiter, 77, 78, 98, 113, 114, 163, 183, 220

Krakatoa, 121, 132, 133, 143, 155, 193, 194

latitude, 34, 37, 43, 44, 46, 106, 157, 172, 206
Laurentide, 41–44, 69, 111
lava, 38, 157, 158
Lichtenberg, Georg Christoph, 81, 85, 91

Magdalenian, 2
magma, 13, 45, 90, 172
magnesium, 89, 159–163
magnitude, 2, 7, 25, 32, 53, 54, 60, 69, 72, 77, 97, 118, 123, 124, 133, 135, 155, 160, 165, 173, 194, 209, 217, 227
mammal, 4, 7, 49, 61, 231, 233
mammoth, iii, 2–5, 8, 21–23, 25, 49, 50, 53, 61
mantle, 45, 46, 99, 171, 235
map, 3, 8, 37, 42–44, 56, 62, 72, 74, 75, 85, 90–92, 105–108, 127
Mars, iii, 67, 68, 71–76, 81, 83, 85, 87, 89–91, 95, 97–101, 105, 109, 110, 112–115, 118, 161, 178, 179, 191
mass extinction, 2, 13, 15, 17, 123, 126, 209, 229–233, 236
Maya, 211–213, 216, 220
mercury, 139–141, 150
mesopause, 33
Mesopotamia, 115, 128, 173, 205, 206, 213–215
meteorite, 11–13, 19, 89–91, 95, 105, 125, 141, 150, 151, 179, 209, 223–225, 228, 233, 234
methane, 120, 121, 123, 180–182, 185, 186, 189, 195, 203
Moon, 79, 80, 98, 141, 160
moon, 77, 78, 103, 160
moraine, 105, 111, 112

nakhlite, 90, 91, 105
NASA, 29, 34, 43, 61, 74, 75, 78, 85, 89, 90, 93, 99–101, 133, 161, 168, 177, 179, 180, 199, 209, 233, 234
neolithic, 63, 144, 145, 205, 207, 208
New Testament, 213
nitrate, 148, 149, 159–161, 163

ocean, 20, 37, 41–43, 45, 46, 56, 59–62, 68, 69, 71–75, 87, 90, 93, 95, 105, 110, 113, 120, 126, 149, 151, 152, 157, 158, 195
Old Testament, 114, 115, 137
Older Dryas, 7, 119, 120, 126
orbit, iii, 46, 74, 80, 89, 97–99, 101, 113, 114, 127–129, 141, 161, 162, 178, 183, 211, 214, 220–222, 227
Orgueil [meteorite], 141, 209
Orontius Finaeus, 105, 108
oxygen, 53, 54, 80, 110, 119, 191, 209

period, 2, 7, 8, 11, 13, 33, 37, 38, 65, 68, 89, 105, 110, 114, 119, 126, 127, 129, 130, 133, 134, 141, 143, 144, 155, 157, 159, 161, 165, 166, 178, 183, 187, 195, 198, 201–203, 205, 206, 208, 211–213, 216, 220, 222, 224, 225, 227, 230, 231
permafrost, 34, 35, 46, 50, 56
Piri Reis, 105, 106
plasma, 56, 77, 78, 80, 102, 169, 181
platinum, 13, 15, 158
Pleistocene, 3, 7, 38, 41, 43, 49, 50, 60, 61, 111, 119, 123, 152, 153, 206
plume, 29, 30, 32, 53, 56, 78, 99, 122, 141, 180
polar, iii, 3, 5, 7, 42, 68, 71, 74, 81, 83, 107, 141, 191
potassium, 11, 15, 101
probe, 71, 78, 86, 99, 100, 102, 162, 180, 181
proton, 165, 166, 169, 191
pyrocumulonimbus, 26

rain, 15, 54–56, 124, 135, 137, 200
river, 21, 46, 49, 50, 69, 145, 199, 201
rock, 12, 19, 26, 37, 38, 41, 45, 46, 54, 83, 87, 89–91, 93, 95, 108, 122, 125, 145, 179, 194, 222, 227
Ross Sea, 43, 108
rotation, 37, 46, 100, 101, 171, 172, 183

scarring [electric], 81, 83, 87, 91, 95
sea level, iii, 59–61, 65, 68, 69, 71, 93, 105, 108–110, 112, 114, 126, 200, 201
sebaceous glands, 4
sediment, 8, 13, 15, 19, 43, 44, 50, 54, 56, 60, 90, 108, 109, 114, 119, 120, 122, 230
seismicity, 19, 73, 137, 163, 165, 168, 169, 171, 172, 181
Shuvalov, Valery, 19, 29, 30, 53
Siberia, 3, 5, 7, 8, 22, 23, 26, 34, 37, 42–44, 46, 49–51, 53, 54, 56, 63, 149, 230
silt, 50
slippage, 7, 45–47, 54, 171, 172
snow, 4, 19, 49, 54, 57, 65, 93–95
soot, 11, 15, 54, 230
speed, 29, 30, 34, 56, 172, 191

spherule, 14, 15, 49, 158, 224
spike, 13, 121–123, 131–133, 135, 139, 141, 143, 144, 148, 149, 151, 152, 155–159, 161–163, 165, 168, 170, 173, 181, 183, 185–187, 189, 192–195, 197, 198, 203, 211, 220, 221, 223
star, 53, 129, 211, 212
Stonehenge, 44
storm, 2, 25, 26, 75, 171
strata, 11–13, 15, 152, 230
stratosphere, 13, 26, 160
subduction, 60, 73
sulfate, 121, 131–133, 141, 143, 144, 159
sulfur, 81, 121–123, 133, 135, 141, 144, 155, 158, 159, 161, 163, 180, 181, 193, 194, 203, 223
Sumer, 115, 127, 128, 173, 178, 211, 212, 215, 216, 222
Sun, 53, 83, 98, 101, 102, 113, 114, 126, 127, 129, 160, 161, 165, 168–170, 178, 192, 211, 212, 221, 227

tablet, 115, 128, 129, 173, 212, 215, 216
tail, iv, 5, 17, 22, 63, 77, 79, 80, 97, 102, 103, 108, 110, 121, 125, 126, 143, 158, 169, 177, 179, 181, 191–193, 197, 207, 208, 214, 217, 224
Tambora, 159
tectonic, 73, 166, 168, 170, 172
telescope, 46, 71, 77
temperate, 3, 5, 23, 25, 26, 34, 43, 44, 53, 185, 203
Teotihuacan, 44
tephra, 134, 144, 155, 158, 194
Thera, 132–135, 143, 158
titanium, 12, 139, 140, 150, 151, 158
tree, 3, 22, 50, 53, 54, 57, 132, 134, 135, 157, 187–189, 200, 229–231
troposphere, 26
tsunami, 15, 59, 60, 114, 124, 222, 224
Turkey, 63, 144

Valles Marineris, 85–87, 89, 91, 95, 105
Velikovsky, Immanuel, iii, 98–101, 114, 115, 180, 181, 183, 211, 213
Venus, iii, 98–103, 113, 114, 177–183, 189, 191–193, 195, 197, 203, 211–216, 220, 227
Vesuvius, 133–135
void, 33, 53, 54
volcano, iii, 26, 30, 37, 54–57, 78, 90, 117, 121, 132–137, 139, 141, 142, 144, 154–161, 163, 165, 166, 169–172, 174, 180, 181, 185, 187, 193, 194, 197, 223, 225
Vostok, 25, 94, 120, 124, 147

wake, 29, 32, 53
water, iii, 4, 5, 20, 30, 37, 43, 44, 54–56, 59, 62, 65, 67–69, 71–74, 83, 85–87, 89, 91, 95, 105,

108–110, 114, 115, 118, 122, 125, 126, 133, 151, 152, 185, 191, 201, 202, 231
wind, 25, 27, 32, 34, 50, 53, 54, 56, 57, 73, 80, 124, 166, 170, 187, 191, 198, 212, 236

yedoma, 50, 51, 56, 57
Younger Dryas, iii, 5, 7, 11–15, 17, 38, 41, 43, 44, 46, 59–63, 65, 68, 69, 72, 105, 109–112, 114, 118, 119, 123, 149, 158, 159

Other books and DVDs from Red Pill Press

"From Paul To Mark"

PaleoChristianity

Laura Knight-Jadczyk

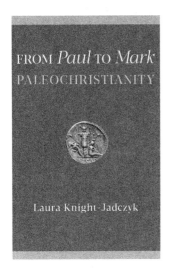

Nearly two thousand years ago the seeds of a new religion were sown in the eastern fringes of the Roman empire. An apostle named Paul wrote letters to his small congregations offering support, rebukes, and the outline of the gospel that would come to be known as Christianity. In the decades after came the Gospel of Mark, followed by more letters and more Gospels, controversies and debates, factions and infighting, until finally, Christianity became an empire.

But what if nearly everything you thought you knew about early Christianity was wrong? When read without preconceptions, the available contemporary sources tell a very different story, filled with 'colorful' characters, hardened revolutionaries, political maneuvering, and ideological conflict. In this groundbreaking study, Laura Knight-Jadczyk strips away centuries of assumptions and dogma to reexamine the fundamentals of what we can truly know of the early Christians, how we know it, and how that changes our picture of what was really happening in first-century Judea.

Why are there no historical references to Jesus and Christianity until decades after the events of the Gospels were supposed to have occurred? Why do the first non-Christian historians who mention Jesus seem dependent on the Gospels? Why does Paul make no unambiguous references to the Gospels' Jesus of Nazareth? What was Paul talking about? Laura Knight-Jadczyk's answers to these questions are revolutionary. After reading this book, you'll never see the origins of Christianity the same way again.

"What will happen to you if you read this book? I'll be glad to tell you. Your paradigm will begin to shift, perhaps only gradually at first. Your assumptions, even your axioms, will be challenged, and this time you will no longer be able to nervously default to the familiar. And all this will happen because you will be seeing the emergence of an exciting new stage of biblical criticism. Laura Knight-Jadczyk has here synthesized the work of a new generation of scholars who are not afraid to venture beyond convention and consensus. She has shown that the work of Wells, Doherty, Doughty, Carrier, Detering, Pervo, and myself are not merely isolated fireworks displays but rather gleams of a new, rising dawn. And in that light she presses on to her own striking advances. Won't you join her?"—Robert M. Price, host of The Bible Geek podcast, author of *Jesus Christ Superstition* and *The Amazing Colossal Apostle*.

"Quite a delight, well written, well researched."—Russell Gmirkin, author of *Plato and the Creation of the Hebrew Bible* and *Berossus and Genesis, Manetho and Exodus*

ISBN 978-1-7349074-1-4

"Earth Changes and the Human-Cosmic Connection"
Pierre Lescaudron with Laura Knight-Jadczyk

Jet Stream meanderings, Gulf Stream slow-downs, hurricanes, earthquakes, volcanic eruptions, meteor fireballs, tornadoes, deluges, sinkholes, and noctilucent clouds have been on the rise since the turn of the century. Have proponents of man-made global warming been proven correct, or is something else, something much bigger, happening on our planet?

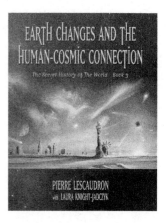

While mainstream science depicts these Earth changes as unrelated, Pierre Lescaudron applies findings from the Electric Universe paradigm and plasma physics to suggest that they might in fact be intimately related, and stem from a single common cause: the close approach of our Sun's 'twin' and an accompanying cometary swarm.

Citing historical records, the author reveals a strong correlation between periods of authoritarian oppression with catastrophic and cosmically-induced natural disasters. Referencing metaphysical research and information theory, *Earth Changes and the Human-Cosmic Connection* is a ground-breaking attempt to re-connect modern science with the ancient understanding that the human mind and states of collective human experience can influence cosmic and earthly phenomena.

Covering a broad range of scientific fields, and including over 250 figures and 1,000 sources, *Earth Changes and the Human-Cosmic Connection* is presented in an accessible format for anyone seeking to understand the signs of our times.

ISBN 978-1-7349074-3-8

"The Secret History of the World", Book 1

... and how to get out alive

Laura Knight-Jadczyk

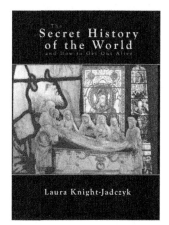

If you heard the Truth, would you believe it? Ancient civilisations. Hyperdimensional realities. DNA changes. Bible conspiracies. What are the realities? What is disinformation?

The Secret History of The World and How To Get Out Alive is the definitive book of the real answers where Truth is more fantastic than fiction. Laura Knight-Jadczyk, wife of internationally known theoretical physicist, Arkadiusz Jadczyk, an expert in hyperdimensional physics, draws on science and mysticism to pierce the veil of reality. Due to the many threats on her life from agents and agencies known and unknown, Laura left the United States to live in France, where she is working closely with Patrick Rivière, student of Eugene Canseliet, the only disciple of the legendary alchemist Fulcanelli.

With sparkling humour and wisdom, she picks up where Fulcanelli left off, sharing over thirty years of research to reveal, for the first time, The Great Work and the esoteric Science of the Ancients in terms accessible to scholar and layperson alike.

Conspiracies have existed since the time of Cain and Abel. Facts of history have been altered to support the illusion. The question today is whether a sufficient number of people will see through the deceptions, thus creating a counter-force for positive change - the gold of humanity - during the upcoming times of Macro-Cosmic Quantum Shift. Laura argues convincingly, based on the revelations of the deepest of esoteric secrets, that the present is a time of potential transition, an extraordinary opportunity for individual and collective renewal: a quantum shift of awareness and perception which could see the birth of true creativity in the fields of science, art and spirituality. *The Secret History of the World* allows us to redefine our interpretation of the universe, history, and culture and to thereby navigate a path through this darkness. In this way, Laura Knight-Jadczyk shows us how we may extend the possibilities for all our different futures in literal terms.

With over 850 pages of fascinating reading, *The Secret History of The World and How to Get Out Alive* is rapidly being acknowledged as a classic with profound implications for the destiny of the human race. With painstakingly researched facts and figures, the author overturns long-held conventional ideas on religion, philosophy, Grail legends, science, and alchemy, presenting a cohesive narrative pointing to the existence of an ancient techno-spirituality of the Golden Age which included a mastery of space and time: the Holy Grail, the Philosopher's Stone, the True Process of Ascension. Laura provides the evidence for the advanced level of scientific and metaphysical wisdom possessed by the greatest of lost ancient civilizations - a culture so advanced that none of the trappings of civilization as we know it were needed, explaining why there is no 'evidence' of civilization as we know it left to testify to its existence. The author's consummate synthesis reveals the Message in a Bottle reserved for humanity, including the Cosmology and Mysticism of mankind Before the Fall when, as the ancient texts tell us, man walked and talked with the gods. Laura shows us that the upcoming shift is that point in the vast cosmological cycle when mankind - or at least a portion of mankind - has the opportunity to regain his standing as The Child of the King in the Golden Age.

If ever there was a book that can answer the questions of those who are seeking Truth in the spiritual wilderness of this world, then surely *The Secret History of the World and How to Get Out Alive* is it.

ISBN 9781897244166

Comets and the Horns of Moses

"The Secret History of the World", Book 2
Laura Knight-Jadczyk

The Laura Knight-Jadczyk's series, The Secret History of the World, is one of the most ambitious projects ever undertaken to provide a cogent, comprehensive account of humanity's true history and place in the cosmos. Following the great unifying vision of the Stoic Posidonius, Laura weaves together the study of history, mythology, religion, psychology and physics, revealing a view of the world that is both rational and breathtaking in its all-encompassing scope. This second volume, Comets and the Horns of Moses, (written in concert with several following volumes soon to be released) picks up the dangling threads of volume one with an analysis of the Biblical character of Moses – his possible true history and nature – and the cyclical nature of cosmic catastrophes in Earth's history.

Laura skillfully tracks the science of comets, revealing evidence for the fundamentally electrical and electromagnetic nature of these celestial bodies and how they have repeatedly wreaked havoc and destruction on our planet over the course of human history. Even more startling however, is the evidence that comets and cometary fragments have played a central role in the formation of human myth and legend and the very concept of a 'god'. As she expertly navigates her way through the labyrinth of history, Laura uncovers the secret knowledge of comets that has been hidden in the great myths, ancient astronomy (and astrology) and the works of the Greek philosophers. Concluding with a look at the political and psychological implications of cyclical cometary catastrophes and what they portend for humanity today, Comets and the Horns of Moses is a marvel of original thought and keen detective work that will rock the foundations of your understanding of the world you live in, and no doubt ruffle the feathers of the many academics who still cling to an outdated and blinkered view of history.

ISBN: 9781897244838

The Wave 1 – "Riding the Wave"

... The Truth and Lies about 2012 and Global Transformation

Laura Knight-Jadczyk

As 2012 fast approaches, opinions about what to expect on this much-anticipated date are sharply polarized. Will humanity experience a global, spiritual transformation? Cataclysmic Earth Changes? Or both? Or nothing? If Earth and its inhabitants are scheduled for some life-changing or life-ending event, we should ask ourselves what we know and how we know it, and how to prepare for our future.

Drawing on decades of research into history, religion, and the esoteric, Laura Knight-Jadczyk introduces the concept of "the Wave" to describe the possible phenomena behind all the hype surrounding global transformation. *Riding the Wave* not only collects the most probable scenarios we may face in the near future – it provides the context to make it all intelligible.

With roots in the science of hyperdimensions made popular by physicist Michio Kaku and the Fortean theories of the late John Keel, *Riding the Wave* suggests that many of the noticeable changes to our world in the last century are symptoms of the approaching Wave. From climate change, extreme population growth and technological development, as well as novel social and political movements, to the advent of UFO sightings, crop circles, and a variety of otherworldly experiences, something is up on the Big Blue Marble, and it all seems to be leading to a sea change in the way we see and interact with the world. The only question is, will it be for the better or the worse?

An intimate blend of science and mysticism, this volume of Laura Knight-Jadczyk's Wave Series initiates the process of unveiling the truth about life on Earth, and the man behind the curtain...

ISBN: 9781897244500

The Wave 2 – "Soul Hackers"

... The Hidden Hands Behind the New Age Movement
Laura Knight-Jadczyk

Why are we here? Why do we suffer? If this is an infinite school, what are we here to learn? And why do our efforts at "fixing" our lives often do exactly the opposite? As mystic and researcher Laura Knight-Jadczyk writes in this volume of her expansive Wave Series: "when you ask a question – if the question is a burning one – your life becomes the answer. All of your experiences and interactions and so forth shape themselves around the core of the answer that you are seeking in your soul. In [my] case, the question was: 'How to be One with God,' and the answer was, 'Love is the answer, but you have to have knowledge to know what Love really is.'"

Soul Hackers is a deeply personal and insightful account of this very process – of burning questions and transformative answers. Through the story of her own struggle with mainstream and alternative religion and the solutions they claim to offer, Knight-Jadczyk lays bare the problems inherent in the New Age movement as a whole – from Reiki, Wicca, and the phenomenon of channeling, to the very real problems of spirit attachments, mind control, and otherworldly predators posing as benevolent beings. She asks what it really means to "create your own reality." Is it merely self-hypnosis, or is something more hidden in this New Age truism?

The answers lie in the very nature of the Wave – the cosmic force and fabric of personal and collective evolution. For anyone wishing to understand the deeper meaning and reality of the human experience, and what our very near future may very well have in store for us, *Soul Hackers* provides a map to our symbolic reality and the knowledge necessary to weather the approaching storm.

ISBN: 9781897244517

The Wave 3 – "Stripped to the Bone"

... The Path to Freedom in the Prison of Life

Laura Knight-Jadczyk

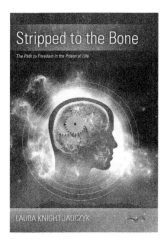

Media propaganda. Official cover-ups. Dishonest science. "Non-lethal" weaponry. Mind control technology. Racial stereotypes. Social engineering. Religious programming. The cold pursuit of profit. And the unrelenting pull of materialism... In a world where "freedom" is exported at the barrel of a gun, true freedom seems more like a distant fairytale, blocked for us in more ways than we can imagine.

In *Stripped to the Bone*, author Laura Knight-Jadczyk lays bare the forces seeking to keep humanity in a prison of its own creation. She lucidly describes evil's place in the cosmos, from the dark world of political conspiracy and government mind control to the reality behind the UFO phenomenon. But in response to the grim state of affairs on the Big Blue Marble, she also asks: Is there a solution? What can we learn from those who came before us? *Stripped to the Bone* suggests that this knowledge was not only known and widely practiced in humanity's prehistory, but that it can be rediscovered.

Through her extensive reading on all things esoteric, Knight-Jadczyk maintains that by knowing our limitations, we may overcome them. In this volume of her acclaimed Wave Series, she tears down our illusions about freedom and the idea that it can be won in any war. Rather, the path to freedom is an inner battle against the many limitations placed on our ability to choose by official culture, our own beliefs, and the forces behind the reality of our everyday experience. By showing us our own limitations she also succeeds in presenting anew the real possibilities and true potential of a free humanity.

ISBN: 9781897244524

The Wave 4 – "Through A Glass Darkly"

... Hidden Masters, Secret Agendas and a Tradition Unveiled

Laura Knight-Jadczyk

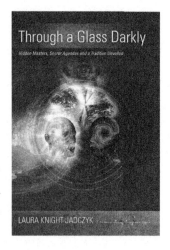

Behind the surface of everyday life lie secrets that have been kept from the eyes of the humanity. In every field of knowledge, we seem to take a wrong turn, coming to conclusions that are diametrically opposed to the truth of the matter. It seems that true science, history, the purpose and aim of human life, our past and potential futures are all off limits to public consumption. How can this be the case, and can these truths come to be known?

In *Through a Glass Darkly*, Laura Knight-Jadczyk continues to make it clear that nothing is what it appears to be. From the stories stitched together to make up our own personal identities to the myths of history on which nations are founded, we live in a sea of lies and half-truths. Just as we lie to ourselves and each other about who we really are, often putting ourselves in the best light possible, there are those who manufacture, manipulate, and shape current and past events to suit their own vested interests. And the current events of today will become the history of tomorrow, erroneously shaping our notions of who we are as a people, just as those of the past have done before.

But behind this sorry state of affairs, the truth awaits discovery. In this fourth volume of her series *The Wave or Adventures with Cassiopaea*, Knight-Jadczyk follows the trail of the hidden masters of our planet, exposing the agenda behind the alleged secret society, the "Priory of Sion", and that mystery's connections with alchemy, Oak Island, and the Kabbalists of old. In the process she reveals aspects of the tradition kept under wraps by these very groups. By exposing the agendas and conspiracies of the elite, we can come to know the truth about ourselves, and why it is has been kept hidden.

ISBN: 9781897244562

The Wave 5 & 6 – "Petty Tyrants & Facing the Unknown"
... Navigating the Traps and Diversions of Life in the Matrix
Laura Knight-Jadczyk

From the myths of romance to the tales of the hero's journey, the quest for knowledge and being has always been portrayed in terms of struggle. Far from home, the hero faces obstacles and tests of his or her courage, will, and cunning. But how do the labyrinths and monsters of these 'messages in a bottle' from our remote ancestors relate to our lives in the 21st century? In an age of mass media, the worldwide web, and multinational corporations, how do these archetypal dramas play themselves out?

In these two volumes of her revolutionary series, *The Wave or Adventures with Cassiopaea*, Laura Knight-Jadczyk continues her project of laying bare the nature of our reality. Through her own experiences and interactions over the course of the Cassiopaean Experiment, many of which just go to show that truth is stranger than fiction, Laura describes the real-life dynamics only hinted at in myth. Most importantly, she gives the tools and clues necessary to actually read the symbols of reality: the theological substrate in which our ordinary psychological motivations are embedded.

With these stunning revelations, Shakespeare's famous words take on a whole new meaning: "All the world's a stage, and all men and women merely players."

First published on her groundbreaking website cassiopaea.org, *Petty Tyrants & Facing the Unknown* have now been fully revised and packaged together in one attractive volume. For anyone interested in the world of esoteric knowledge, studies in the paranormal and everything 'alternative', or even just curious about life in general and its possible significance and meaning, these volumes are a must-read.

ISBN: 9781897244630

The Wave 7 – "Almost Human"

... A Stunning Look at the Metaphysics of Evil
Laura Knight-Jadczyk

In this volume of her prescient Wave series, Laura Knight-Jadczyk brings order to the chaotic and labyrinthine world of murder, conspiracy, and the paranormal. In a unique and probing synthesis of science and mysticism she presents a detailed series of case studies and application of her hypothesis of hyperdimensional influence.

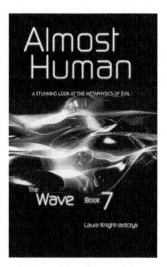

From interpersonal relationships and their expression of archetypal dramas to the vectoring of human behavior to achieve hyperdimensional purposes, *Almost Human* reveals the mechanics of evil, how it creeps into our lives, and what we need to be aware of in order to avoid it.

The case studies of John Nash, the schizoidal creator of Game Theory, and Ira Einhorn, the New Age psychopath who murdered his girlfriend, are the window through which Knight-Jadczyk unravels the intricate web of deception, aims, and counter-aims of the Powers That Be.

Almost Human is essential reading for anyone wondering why our world is becoming increasingly controlled and our freedoms more restricted.

ISBN: 9781897244463

The Wave 8 – "Debugging the Universe"

... The Hero's Journey

Laura Knight-Jadczyk

The Path of the Fool, the Hero's Journey, the Great Work – by whatever name it takes, the path of self-development and growth of knowledge is one fraught with difficult lessons and intense struggle. But what exactly is the nature of those lessons, and what insights can the latest advances in modern science provide for us along the way?

Debugging the Universe takes us into the heart of what it means to be human, from the molecules of our DNA to our life purpose and true place in the universe, and everything that separates us from embodying that higher potential. Explored within are real-life applications of the Hero's archetype, the relevance of neuroscience and the 'molecules of emotion', the hidden meaning behind the enigmatic symbols of esoterica, and what it means to live inside a complex system: the universal breath of chaos and order.

This volume concludes the publication in print of Laura Knight-Jadczyk's unparalleled and controversial magnum opus: *The Wave or Adventures with Cassiopaea*. Originally published online at www.cassiopaea.org, The Wave is a fully modern exposition of the knowledge of the ancients, with subjects ranging from metaphysics, science, cosmology, and psychology to the paranormal, UFOs, hyperdimensions and macrocosmic transformation.

ISBN: 9781897244654

"High Strangeness"

... Hyperdimensions and the Process of Alien Abduction

Laura Knight-Jadczyk

High Strangeness: Hyperdimensions and the Process of Alien Abduction is an enlightening attempt to weave together the contradictory threads of religion, science, history, alien abduction, and the true nature of political conspiracies. With thorough research and a drive for the truth, Laura Knight-Jadczyk strips away the facades of official culture and opens doors to understanding our reality.

The Second Edition includes additional material that explains the hyperdimensional mechanisms by which our reality is controlled and shaped by 'alien' powers. The self-serving actions of unwitting puppets – psychopaths and other pathological types – who may have no knowledge that they are being used, become the portals through which an agenda that is hostile to humanity as a whole, is pushed forward.

High Strangeness takes the study of ponerology into a whole new dimension!

ISBN: 9781897244340

"The Cassiopaea Experiment Transcripts"

Volume 1, 1994

Laura Knight-Jadczyk

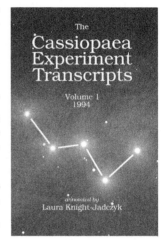

The Cassiopaea Experiment is unique in the history of channeling, mediumship, and parapsychology. For years prior to the first Cassiopaean transmission, Laura Knight-Jadczyk went to great lengths to study the channeling phenomenon, including its history, its inherent strengths, weaknesses, dangers, and the various theories and methods developed in the past. After having exhausted the standard literature in search of answers to the fundamental problems of humanity, Laura and her colleagues (including her husband, mathematical physicist Arkadiusz Jadczyk) have held regular sittings for more than twenty years.

With the goal of applying true scientific standards and critical thinking, Laura began her experimentation with the spirit board, chosen for the optimum conditions of conscious feedback it offers. After two years of working through various levels of phenomena, including alleged discarnate entities and denizens of the 'astral realms', a new source came through in 1994. Unlike previous contacts, the Cassiopaeans came through strong and clear, often delivering full paragraphs' worth of complex material at one to two letters per second, with a high density of content and impeccable orthography. After telling Laura, "we are you in the future," the C's have been covering a wide range of topics by transmitting more than one million letters over the last twenty years, answering questions from practically all fields of knowledge, including physics, mathematics, psychology, philosophy, parapsychology, esotericism, history, politics, health, and astronomy. Where the material can be verified, the C's have proven to have an amazing track record, especially when compared to the variable quality of most examples of channeling from the past 150 years.

For the first time in print, this volume includes complete and extensively annotated transcripts of 36 experimental sessions conducted in 1994, beginning with the first contact with the Cassiopaeans on July 16. The sessions of this year introduced many of the themes that would recur in more detail over the next twenty years, including such topics as cyclical cometary bombardment of the Earth, the solar companion hypothesis, ancient history, metaphysics, the hyperdimensional nature of reality, and the possibility of evolution of humanity.

ISBN: 978-1-897244-99-9

"The Cassiopaea Experiment Transcripts"

Volume 2, 1995

Laura Knight-Jadczyk

The Cassiopaea Experiment is unique in the history of channeling, mediumship, and parapsychology. For years prior to the first Cassiopaean transmission, Laura Knight-Jadczyk went to great lengths to study the channeling phenomenon, including its history, its inherent strengths, weaknesses, dangers, and the various theories and methods developed in the past. After having exhausted the standard literature in search of answers to the fundamental problems of humanity, Laura and her colleagues (including her husband, mathematical physicist Arkadiusz Jadczyk) have held regular sittings for more than twenty years.

With the goal of applying true scientific standards and critical thinking, Laura began her experimentation with the spirit board, chosen for the optimum conditions of conscious feedback it offers. After two years of working through various levels of phenomena, including alleged discarnate entities and denizens of the 'astral realms', a new source came through in 1994. Unlike previous contacts, the Cassiopaeans came through strong and clear, often delivering full paragraphs' worth of complex material at one to two letters per second, with a high density of content and impeccable orthography. After telling Laura, "we are you in the future," the C's have been covering a wide range of topics by transmitting more than one million letters over the last twenty years, answering questions from practically all fields of knowledge, including physics, mathematics, psychology, philosophy, parapsychology, esotericism, history, politics, health, and astronomy. Where the material can be verified, the C's have proven to have an amazing track record, especially when compared to the variable quality of most examples of channeling from the past 150 years.

For the first time in print, this second volume includes complete transcripts of 51 experimental sessions conducted in 1995. Questions and answers have been annotated extensively, giving unprecedented insight into the background and interpersonal dynamics of the early Cassiopaea Experiment.

In this year, the dialog with the Cassiopaeans revolved around topics of 'New Age' and historical disinformation, reincarnation and past lives, karma, soul evolution and the purpose of humanity, the 'Golden Age' and ancient civilizations, weather and Earth changes, 'The Wave' and coming turmoil on our planet, the 'New World Order', hidden government and the all-encompassing control system, higher 'density' experience and the 'fluid' nature of the UFO and alien phenomenon, time and the nature of our universe. Also, interspersed with these topics, the sessions of this year document the conflicting interests arising within the group undertaking the Cassiopaea Experiment and the struggle against sometimes obvious skewing of the information and attempted sidetracking of the communications.

ISBN: 978-0-692484-51-7

"The Cassiopaea Experiment Transcripts"
Volume 3, 1996
Laura Knight-Jadczyk

The Cassiopaea Experiment is unique in the history of channeling, mediumship, and parapsychology. For years prior to the first Cassiopaean transmission, Laura Knight-Jadczyk went to great lengths to study the channeling phenomenon, including its history, its inherent strengths, weaknesses, dangers, and the various theories and methods developed in the past. After having exhausted the standard literature in search of answers to the fundamental problems of humanity, Laura and her colleagues (including her husband, mathematical physicist Arkadiusz Jadczyk) have held regular sittings for more than twenty years.

With the goal of applying true scientific standards and critical thinking, Laura began her experimentation with the spirit board, chosen for the optimum conditions of conscious feedback it offers. As Laura writes, "We take the approach of a sort of scientific mysticism where mystical claims are submitted to rational analysis and testing, and the required scientific proofs are modified to allow for the nature of evidence from theorized realms outside of our own where ordinary scientific proofs might not apply."

The dynamics and content of channeling are in general tightly entangled with current global events and the lives of the participants and their interpersonal relationships. For this reason, clarifying context has been added, and questions and answers have been annotated extensively, giving unprecedented insight into the background of the Cassiopaea Experiment.

In this year, the dialogue with the Cassiopaeans revolved around the further exploration of various theories that were promulgated at that time in books and the internet, the solar companion hypothesis, the theorized states of being above our own, hypnosis, mind programming, false prophecies, the rediscovery of a possible ancient technology utilized in the pre-history of mankind, quantum principles and possibilities for their macroscopic application, as well as the process of gaining more protection by assimilating and applying knowledge and networking with like-minded people.

For the first time in print, this volume includes complete transcripts of 44 experimental sessions conducted in 1996.

ISBN: 978-0-692725-73-3

"The Cassiopaea Experiment Transcripts"

Volume 4, January – May 1997

Laura Knight-Jadczyk

Following the various clues given by the Cassiopaeans, Knight-Jadczyk includes email correspondence documenting the early stages of her research leading to the publication of her book *The Secret History of the World*, as well as illuminating the rationale behind the questions and answers in this year – a year that demonstrates the material manifestations of how reality shifts begin.

For the first time in print, this volume includes complete transcripts of 17 experimental sessions conducted between January and May 1997.

The Cassiopaea Experiment is unique in the history of channeling, mediumship, and parapsychology. For years prior to the first Cassiopaean transmission, Laura Knight-Jadczyk went to great lengths to study the channeling phenomenon, including its history, its inherent strengths, weaknesses, dangers, and the various theories and methods developed in the past. After having exhausted the standard literature in search of answers to the fundamental problems of humanity, Laura and her colleagues (including her husband, mathematical physicist Arkadiusz Jadczyk) have held regular sittings for more than twenty years. With the goal of applying true scientific standards and critical thinking, Laura began her experimentation with the spirit board, chosen for the optimum conditions of conscious feedback it offers. As Laura writes, "We take the approach of a sort of scientific mysticism where mystical claims are submitted to rational analysis and testing, and the required scientific proofs are modified to allow for the nature of evidence from theorized realms outside of our own where ordinary scientific proofs might not apply."

ISBN: 978-1-548314-87-3

"9/11 – The Ultimate Truth"

Joe Quinn and Laura Knight-Jadczyk

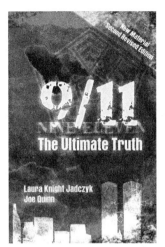

In the years since the 9/11 attacks, dozens of books have sought to explore the truth behind the official version of events that day - yet to date, none of these publications has provided a satisfactory answer as to **why** the attacks occurred and who was ultimately responsible for carrying them out.

Taking a broad, millennia-long perspective, Laura Knight-Jadczyk's *9/11: The Ultimate Truth* uncovers the true nature of the ruling elite on our planet and presents new and ground-breaking insights into just how the 9/11 attacks played out.

9/11: The Ultimate Truth makes a strong case for the idea that September 11, 2001 marked the moment when our planet entered the final phase of a diabolical plan that has been many, many years in the making. It is a plan developed and nurtured by successive generations of ruthless individuals who relentlessly exploit the negative aspects of basic human nature to entrap humanity as a whole in endless wars and suffering in order to keep us confused and distracted to the reality of the man behind the curtain.

Drawing on historical and genealogical sources, Knight-Jadczyk eloquently links the 9/11 event to the modern-day Israeli-Palestinian conflict. She also cites the clear evidence that our planet undergoes periodic natural cataclysms, a cycle that has arguably brought humanity to the brink of destruction in the present day.

For its no nonsense style in cutting to the core of the issue and its sheer audacity in refusing to be swayed or distracted by the morass of disinformation that has been employed by the powers that be to cover their tracks, *9/11: The Ultimate Truth* can rightly claim to be **the** definitive book on 9/11 - and what that fateful day's true implications are for the future of mankind.

The new Second Edition of *9/11: The Ultimate Truth* has been updated with new material detailing the real reasons for the collapse of the World Trade Center towers, the central role played by agents of the state of Israel in the attacks, and how the arrogant Bush government is now forced to dance to the Zionists' tune.

ISBN: 9781897244876

JFK: The Assassination of America

Laura Knight-Jadczyk

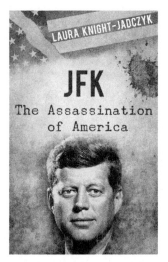

Anyone who has taken the time to study the facts about that fateful day in Dallas, TX, November 22, 1963, will already know that John F. Kennedy was deliberately murdered by a cabal of psychopathic warmongers who were opposed to his plans for a more peaceful world. This ebook written by Laura Knight-Jadczyk brings into focus how the convergence of greed and the power-mad forces of big oil, organized crime, and the military-industrial complex brought about the destruction of JFK. Drawing on an early analysis of Kennedy's assassination, *Farewell America*, which was produced by a French intelligence group, Mrs. Knight-Jadczyk brings a deeper understanding of this tragic event by placing it in the light of the psychopathic motivations of these criminal elements. *JFK: The Assassination of America* shows a world that could have been, and a great man silenced by forces who will stop at nothing to keep that world from becoming a reality.

ISBN: 9781897244883 (Amazon Kindle only)

Amazing Grace

Laura Knight-Jadczyk

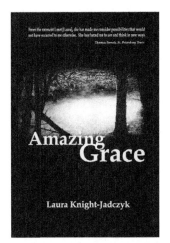

Laura Knight-Jadczyk has lived intimately – and mysteriously – with the world of spirit. In *Amazing Grace*, Laura takes us back to her beginnings in a Gulf Coast Florida childhood, mapping the first decades of her extraordinary search for an objective reality of spirit, of the play of forces that exist as a subtext to the lives of all human beings – a journey toward knowledge and understanding.

From her first experiences with a terrifying Face at the Window in childhood, to her work as an exorcist, chronicled by Pulitzer Prize–winning journalist Thomas French in the *St. Petersburg Times*, Laura relates the many experiences in her search for the existence of truth about our reality, which forced her to recognize the validity of perceptions beyond those of materialism.

This is also the story of how the Cassiopaeans came to be a part of her life. Their channeled messages, which include important concepts of physics and the underlying nature of reality, have drawn the attention of intellectually advanced yet spiritually hungry people from all over the world. This is not just the story of one woman's experience with personal quantum jumps from one reality to another, but is also the greater story of the potential that exists in every seeker. We have the potential to discover the genuine existence of spirit and the play of the archetypal forces of the world, and to connect with them in a dynamic way. Amazing Grace, or Quantum Future, can be a reality in our lives.

ISBN: 9781897244814

The Apocalypse: Comets, Asteroids and Cyclical Catastrophes

Laura Knight-Jadczyk

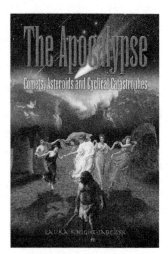

For untold millennia, comets and asteroids have struck fear into the hearts of humankind. Their stark radiance was observed everywhere with a sense of impending doom, interpreted as signs of the gods' judgment, omens of plague, mass destruction and the end of time. Astronomers recorded their appearance the world over, building large scale observatories to track their movements and predict their ominous arrival. What was it about these majestic wonders of the heavens that inspired such dread? Was it simply a product of mere superstition and social hysteria?

The latest scientific analysis and historical analysis strongly suggest otherwise. Our ancestors knew something we have since forgotten, their secrets deeply embedded in the archaeological record and the myths passed on throughout generations. And we have only begun to unravel their mysteries ...

Spurred on by the discovery of a little known letter of warning to the European Office of Aerospace Research and Development by astrophysicist Victor Clube, author Laura Knight-Jadczyk began an in-depth research project to get to the bottom of the very real threat to humanity posed by these celestial visitors. In *The Apocalypse: Comets, Asteroids, and Cyclical Catastrophes*, Knight-Jadczyk shares what she found: historical evidence for mass destructions, comet-borne plagues, and repeated cover ups littering our past, as well as clues that a similar fate may be fast approaching.

ISBN: 9781897244616

The Noah Syndrome

Laura Knight-Jadczyk

"As it was in the days of Noah ..."

Technological progress married with moral decay. A people enraptured by the trivial and superficial, entrenched in a culture of materialism and endless warfare. A civilization whose time has come. If a phrase defines the condition of our era, it is this: The Noah Syndrome.

After twenty-six years, Laura Knight-Jadczyk's unpublished book is now in print for the first time. And it's more relevant than ever. Drawing on prophecies ancient and new - from biblical narratives to modern-day visionaries - yet grounded in cutting-edge scientific discoveries about earth's cataclysmic history, this book presents a remarkable vision of humanity's dramatic past and extremely hazardous future.

The Noah Syndrome also introduces the concept of quantum cosmic metamorphosis - the spiritual ark that may carry us through the coming catastrophe. If our past is the key to our future, as Laura suggests, heeding the counsel in these pages could mean the difference between transformation and destruction.

ISBN: 9781897244791

Personality-Shaping Through Positive Disintegration

Kazimierz Dabrowski

For psychologist and psychiatrist Kazimierz Dabrowski, personality is not a given – it must be consciously created and developed by the individual. In his second English-language book, *Personality-Shaping Through Positive Disintegration*, first published in 1967, Dr. Dabrowski presents a comprehensive treatment of personality that is still relevant, perhaps more so today than when it was first written. Here Dabrowski describes personality's individual and universal characteristics, the methods involved in shaping it, and case studies of famous personalities (including Augustine and Michelangelo) demonstrating the empirical and normative nature of personality development.

Included in this edition are the original introduction, written by former APA president O. Hobart Mowrer, an appendix detailing a study on gifted children and outstanding abilities conducted by Dr. Dabrowski, as well as previously unpublished biographical pieces analyzing the personalities of Beethoven, Kierkegaard, and Unamuno. Grounded in Dabrowski's theory of positive disintegration, Personality-Shaping introduces the concepts at the heart of the theory and at the heart of human potential, creativity, social service, inner conflict, mental illness, and personal growth. Dabrowski's all-embracing perspective is at once a fresh alternative to the one-dimensional theories and trends pervasive in the field of psychology, and a full statement in its own right of all those aspects of human nature too often marginalized, ignored, or denied – a revolutionary and heartfelt product of Dr. Dabrowski's incisive observations and all-embracing vision.

ISBN: 978-0-692427-49-1

Fear of the Abyss

Aleta Edwards, Psy. D.

Most self-help or personal growth books help their readers learn to "cope" better with one specific problem that troubles them. In contrast, *Fear of the Abyss* offers healing for a wounded core by providing guidance to those with a certain constellation of issues, which Dr. Edwards calls the "PCS" personality. While perfectionism, control, and shame may be the issues that readers most easily relate to, they form part of a pattern that also includes such issues as rigidity, black-and-white thinking, indecisiveness, and relationship difficulties caused by a need to be needed.

Dr. Edwards sees this entire constellation as rooted in a core fear that varies from person to person. It is the object of this fear that she calls the "abyss." Through a series of clinical vignettes drawn from her own experience as well as exercises accompanying most chapters, she helps the reader develop self-awareness and compassion for him- or herself, as well as a greater sense of authenticity.

"This book is a true gift. Aleta Edwards is ... a genius in simplifying how to get to the root of the issues that trouble us – how to embrace the shadow and transcend to love, compassion, and honor. *Fear of the Abyss* is a must read for every psychotherapist, not to mention anyone who suffers from shame and perfectionism – which I'd say is most if not all all of us." –Lisa Ortigara Crego, PhD, LCSW, CEDS, CAP, NBCCH, Clinical Psychotherapist, Doctor of Addictions Psychology, Nat. Bd. Cert. Hypnotherapist

"In this engaging book, Dr. Aleta Edwards sheds new light on the issues of perfectionism, control, and shame. Breaking from the traditional model of diagnosis, she explains the role of trauma as a major factor in the development of the 'PCS' personality and emphasizes the adaptive nature of these issues. This book is recommended for its thoughtful text and helpful exercises, as well as the spirit of hope." –Nina Savelle-Rocklin, PsyD, Psychoanalyst, award-winning writer of the "Make Peace with Food" blog and host of the "Win the Diet War" podcast

"This is a great book for individuals who struggle with shame and perfectionism. I've recommended it to several clients and they have found it very helpful. I believe this book is a good tool as an addition to therapy or by itself as a self-help book." –Ana Maria Aluisy, MA, LMFT, LMHC, CRC, relationship specialist, featured on CNN Latino, Super Q Radio and ESPN Deportes Radio

ISBN: 978-0-692717-0-59

Evidence of Revision

Quantum Future Group

Evidence of Revision is a six part documentary containing historical, original news footage revealing that the most seminal events in recent American history have been deeply and purposefully misrepresented to the public. Footage and interviews provide an in-depth exploration of events ranging from the Kennedy assassinations to the Jonestown massacre, and all that lies between.

The footprints left in this archival footage reveal the coordinated, clandestine sculpting of the America we know today. Evidence of Revision proves once and for all that history has been revised even as it was written!

Part 1: The Assassinations of Kennedy and Oswald Part 2: The why of it all referenced to Vietnam and LBG Part 3: LBJ, Hoover and others: what so few know even today Part 4: The RFK assassination as never seen before Part 5: RFK assassination, MKULTRA and the Jonestown massacre Part 6: The assassination of Martin Luther King

6 part, 3 DVD set. Region-Free DVDs – Watch on any DVD player or computer anywhere in the world.

Newly subtitled in English, Spanish, French and Polish – the documentary is filmed in English, the subtitles are for clarity due to archival footage on which the audio is, at times, unclear.

Duration: 10 hours 25 minutes
Cover Design: Humberto Braga
ISBN: 9781897244784

Éiriú Eolas, An Amazing Stress Control, Healing and Rejuvenation Program

Laura Knight-Jadczyk

Are you stressed? Do you suffer from chronic fatigue, conditions that your doctor cannot diagnose or that he thinks are "all in your head"? Are you in physical pain more often than not? Is your system toxified from living in today's polluted environment? Do you wish you could face life's challenges with greater calm and peace of mind? Would you like to actually feel healthy, happy and pain-free every day?

"Introducing Éiriú Eolas (pronounced "AIR-oo OH-lahss"), the amazing scientific stress-control, healing, detoxing and rejuvenation program which is THE KEY that will help you to change your life in a REAL and immediately noticeable way:

Proven benefits of the Éiriú Eolas Program include: instantly control stress in high energy situations, detox your body resulting in pain relief, relax and gently work through past emotional and psychological trauma and regenerate and rejuvenate your body/mind.

Éiriú Eolas will enable you to rapidly and gently access and release layers of mental, emotional and physical toxicity that stand between you and a healthy, younger feeling and younger looking body!"

The Éiriú Eolas technique grew out of research conducted by the Quantum Future Group under the direction of Laura Knight-Jadczyk and Gabriela Segura, M.D. The practice has been thoroughly researched and proven to work by the thousands of people who are already benefiting from this unique program. The effects are cumulative and results and benefits can be seen in only a very short time, sometimes after just one session!

Disc 1, PART 1: In this video lecture, you will receive a detailed overview of the science behind the breath and the simple action of breathing by program developer, Laura Knight-Jadczyk. She outlines the importance of the breath for our physical, emotional, and spiritual well being. You will also learn the primary breathing technique that will be used in the exercises on Discs 1, part 2, and disc 2: Pipe Breathing. This is the Magic Key to instant stress reduction and control. Pipe breathing brings the entire body-mind system into balance. You can use it any place at any time for instantaneous results; control fears and phobias; recover quickly from shocks or use in high energy situations where decisive action is required and stress control is essential.

PART 2: This disc is a video demonstration of the Éiriú Eolas warm-up exercises and the breathing techniques that comprise the Éiriú Eolas program.

Disc 2 is an audio CD for you to use to practice the entire Éiriú Eolas breathing program itself, culminating in the powerful Prayer of the Soul, a non-denominational, non-sectarian, scientific prayer - a truly unique achievement in our time.

Disc 3: The audio from the CD with subtitles, set against a backdrop of relaxing nature scenes. Use this disc if Éiriú Eolas CD is not available in your language or if you are hearing impaired. This disc also contains a copy of the Éiriú Eolas guide book in languages in PDF format that can be accessed from your computer along with MP3 versions of the CD in English and other languages.

Subtitles available in: English, Danish, German, Spanish, Greek, French, Croatian, Italian, Dutch, Polish, Russian, Serbian, Turkish and Vietnamese!

ISBN: 9782916721187

Red Pill Press
info@redpillpress.com
www.redpillpress.com

Printed in Great Britain
by Amazon